The Fighting America Flying Boats of WWI | Volume 2

A Centennial Perspective on Great War Seaplanes

Colin A. Owers

Great War Aviation Centennial Series #23

To the late J. M. Bruce, Scottish gentleman and friend. Jack Bruce stunned the world of WWI aviation historians when he published his *British Aeroplanes 1914–1918*, which remains a classic of its kind more than 50 years after its publication. His unprecedented detailed histories of the aircraft of the conflict set new standards of historical reporting on the technology of the period.

Acknowledgements

To friends old and new. Amy Brennan; Carl Bobrow; the late J.M. Bruce; Mark Evans; Simon Frost; Frits Gerderson; Joe Gertler; Hill Goodspeed; the late Peter M. Grosz; Simon Gunn; Dan Hagedon, Adrian Hellwig; Jack Herris; Colin and Barbara Huston; Sir Peter Jackson; R. Kinter; Peter London; Bob Pearson; Bryan Ribbans; Ray Rimell; Noel Shirley; the later Ian R. Stair, Bill Toohey; the late Harry Woodman; the staffs of the RAF Museum, Hendon; the UK and USA National Archives; the Emil Buehler Library; US National Naval Aviation Museum, Pensacola; the Museum of Flight, Seattle; the Curtiss Museum, Hammondsport; and of the Aviation Heritage Trust of New Zealand; and special thanks to my wife Julie who copied many documents that were used in the compilation of this book.

Thanks to Steve Anderson for the cover painting ***Rendezvous Point***. Please see Steve's website at: **www.anderson-art.com**

Color aircraft profiles © Bob Pearson. While every care is taken, some colors are speculative based on known practices at the time. Purchase Bob's CD of WWI aircraft profiles for $50 US/Canadian, 40 €, or £30, airmail postage included, via Paypal to Bob at: **bpearson@kaien.net**

ISBN: 978-1-935881-45-2

Design and layout: Jack Herris
Cover design: Aaron Weaver
Digital photo editing: Aaron Weaver & Jack Herris

Aeronaut Books

www.aeronautbooks.com

Table of Contents

Saunders F.5 N178 hollow hull experimental.

Below: N178 had an experimental Saunders hollow hull. See P.431 for more photographs of this interesting experiment.

Foreword to Volume 2

The First World War saw the first use of submarines in strength and the early German successes led to the Admiralty becoming aware of the role that aircraft could play in suppressing this menace. While it was the convoy system that eventually led to the defeat of the U-Boat, the part played by aircraft in carrying out bombing attacks on submarines, and by bringing surface craft to the location of a submarine, thus preventing the submersible from carrying out its role, cannot be overlooked. The submarines had to surface and use their diesel engines to charge their batteries and when this became too dangerous to undertake in daylight they were restricted to carrying this out at night. In addition the long-range flying boats performed in the intelligence role with reconnaissance flights to the High Seas Fleet anchorages.

There was a human side to the story and this is brought out in Capt A.E. Boulton's experiences where he records what was required of the crew of such flying machines. The trial of John Porte for corruption was a most remarkable episode in a remarkable story. That he was let off with a virtual "slap on the wrist" indicates how important he and his work in the development of flying boats were regarded.

The Felixstowe flying boats had a long post-war history and much more research needs to be undertaken to document this, particularly with those boats that served in foreign navies.

In Volume 1 the development of the Curtiss and Felixstowe flying boats was traced from their pre-war beginnings to their use in combat. In this volume the story of the other *Small America* boats, the Norman Thompson N.T.4 and N.T.4a, completes the coverage of the *America* boats, while that of the Felixstowe Fury completes the Felixstowe story.

The Curtiss and Felixstowe flying boats were exceptional performers for their day and it is hoped that this volume will complete their story in detail for the aero-historian and modeler.

Colin A Owers

S. E. Saunders-built F.2A N4099 was at Felixstowe by 17 October 1918. It remained there until at least January 1919 after which it went to Calshot. It was written off in a fatal crash in the Solent in 1924.

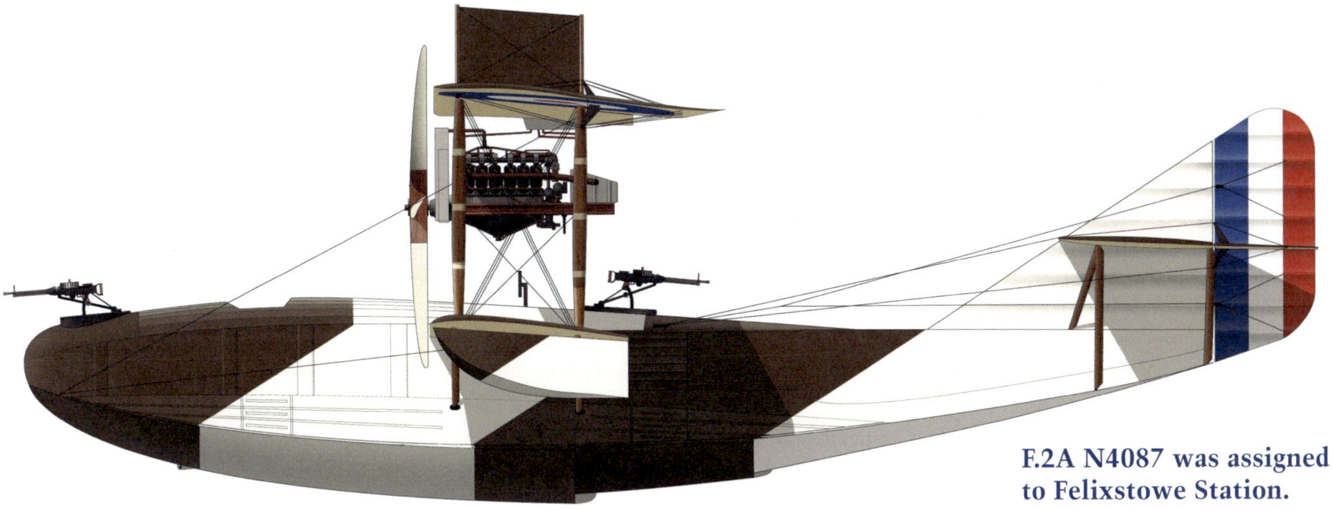

F.2A N4087 was assigned to Felixstowe Station.

6. The America Flying Boats in Detail

Above: One of the original Curtiss H-12 boats in the Curtiss factory. The round radiators are only seen on these early machines with Curtiss engines.

Curtiss Model H-1 and H-4

A Curtiss catalogue for 1915 described the "180–320 H.P. Flying Boat Model H" as "the largest flying boat in the world. It was originally designed to carry two men, their navigating instruments, food and supplies on a trans-Atlantic trip of 1,800 miles with one stop... Many duplicates have since been built by the Curtiss Aeroplane Co. and have given excellent service." Nowhere does it mention for whom the "duplicates" were constructed.

Rather grandly it stated that the machine "will ride safely on the open sea and lift freely from ordinary waves." The forward section of the hull was described as of the "vee-bottom hydroplane type." Aft of this was a "rectangular tail" that terminated in a vertical knife edge where the rudder was attached. Planking was of spruce, the forward section having a double layer, the skins separated by a double layer of Sea Island cotton set in glue. The frames and planking were fastened with brass and copper screws and rivets. Dual controls were fitted. Extension fins, of watertight construction, added to the planning surface of the hull without greatly increasing its bulk.

The wings were in seven panels: a centre section above the engines, four main sections (two upper and two lower) and the upper wing extensions and were constructed of ash and spruce with piano wire internal bracing and covered with unbleached linen. Sub-wings were built into the hull. The wing extensions were braced to king posts above the outer interplane struts. Wing tip floats were mounted under the lower wing. The original *Americas* do not appear to have had dihedral applied, however the British rigging instructions for the H-4 reveal that the panels outside the engine section were installed with 1¾° of dihedral. The tailplane was set neutral until a test flight revealed what position it should be rigged to.

The rudder was of similar construction but the lower portion was constructed of oak and bound with metal for steering in the water. Power was supplied by two 90-hp Curtiss engines, however Curtiss 160-hp V-2 engines could also be installed. The main fuel tanks were in the hull and fuel was

Above: This appears to be 8654. The colouring of the lower hull and fins is clearly seen. This machine does not have the window let into the side for the wireless operator at this time. (via P. London)

pumped to a gravity tank in the upper wing centre section.

The original two Curtiss built *America* flying boats had boat-builder type hulls. The following boats that were designated H-4 by the British followed the same construction methods although those built by the British deviated from the originals. As described in the earlier chapters, the progression of the development of the Felixstowe boats did not proceed well until John Porte abandoned the Curtiss type hull and developed his own method of hull construction that resulted in a lighter, stronger and more easily built hull.

It is not known if any of the A*merica* boats were built in Canada as Canadian histories do not mention the type. It is thought the contracts may well have been given to the Curtiss Company through their Canadian subsidiary in order not to embarrass the US Government that was neutral at this time.

At a Meeting of the Admiralty Air Department on 4 September 1917, it was reported that Felixstowe wanted two 150-hp Hispano Suiza engines for use in a Small America for training purposes. This boat apparently had a special hull but the paper imprint is illegible. Two engines that required slight repairs were made available and sent to Felixstowe. It has not proved possible to identify this machine to date.[1]

Curtiss H-1 & H-4 Small America Flying Boat Orders

Serials	Number	Contract	Manufacturer	Notes
950–951	2	CP52840/14	Curtiss, Hammondsport, NY	Original America trans-Atlantic boats
1228–1235	8	CP65070/14	AMC Ltd	British version of the original *America* boat
1236–1239	4		Curtiss Aeroplane and Motor Co	H-4
3545–3594	50	CP01533/15	Curtiss	H-4

Notes: 1. Curtiss America Specifications on p.34

2. In a 1915 RNAS document under the entry for "America Flying Boats, Improved type," the engines are listed as 2 x 150-hp Sunbeams with the armament quoted as either:

Scouting: 6 pdr Davis Gun with 360 lbs ammunition, and
1 Lewis gun with 80 lbs ammunition, or

Bombing & Coast Patrol: 4 x 100-lb bombs and
6 pdr Davis Gun with 360 lbs ammunition

Above & Facing Page: 8681 was an early H-12 in the two colour hull scheme. In these photographs the two colours have faded and worn. The size and shape of the W/T operator's window just forward of the wings should be noted as these were not on the original hull as supplied by Curtiss but were a station modification. The airscrews appear to have the grey painted canvas covers and metal tips. Although at anchor, 8681 carries a bomb under the starboard wing. 8681 was used for experiments with towed lighters. On 14 January 1918, it suffered a forced alighting and had to be towed back. Apparently the damage it received led to it being deleted later that month.

At the following week's meeting the training of pilots for *Large America* boats was discussed. The machines had proven so valuable that they had been used exclusively for patrols and it had not been possible to allocate any to training. The position now had to be faced or else there would be no pilots for the machines on order. It was stated that the present numbers had been allocated to last out until the end of September and it was hoped that some more machines would be in service by then allowing for machines for training to be allocated in the winter. Cdr Bigsworth said that he could allow three *Large America* boats for training at Calshot but that the C in C insisted on all patrols being carried out on *Large Americas* wherever possible. It was suggested that patrols be carried out on Shorts to free up *Large Americas*. The question was posed whether the full value was being received from *Small America* boats with Anzani engines. There were a few hulls ready for erection at Felixstowe and it was suggested that they be erected as opportunity allowed and used for training pilots prior to their training on the *Large Americas*. "Deliveries from a new contract for *Small Americas* would start in October." It is assumed that this is a reference to the Norman Thompson N.T.4 as it is referred to later at the Meeting as a training machine.[2]

Right: A Curtiss H-12B displays the cockpit canopy and revised bow that distinguishes this version from the earlier H-12 boats. The display of the US flag would indicate that this photograph was taken after the US declared war in 1917.

Above: Curtiss H-12B N4343 was from the batch N4330 to N4353 and displays the changes made to the bow with the gun ring and new cockpit enclosure. This photograph shows much atmosphere on a RNAS station. Note the hangars with their camouflage paint scheme, the floatplane in the background, prop holding the rear of the boat up, the different clothing, especially the sheep skin lined thigh boots on the airman standing on the walkway over the fin. Note the light overall colour (Curtiss English Khaki?) to this boat. The upper wing bears the cockades on upper and lower surfaces. N4343 was stationed at Killingholme when on 13 June 1918, it attacked a submarine that was then sunk by destroyers. This machine also attacked U-boats on 19 and 20 July. Used by the USN, it was still at Killingholme on 30 January 1919.

The Director of Air Services issued a forecast of the proposed position of training machines in December 1917 in which he proposed to send three *Small Americas* "now at Calshot" to either Felixstowe or Killingholme.[3] It was stated that there were only 18 FBA and N.T.2b flying boats at Calshot and 12 FBA boats at Killingholme "an absurdly small bottleneck to pass pilots through" when the requirements were for 150 *Large Americas* and 375 floatplanes, the *Americas* requiring two pilots and the floatplanes one. "A further point that appears to need discussion is D.A.S. ruling that every Large America pilot <u>must</u> have done 20 hours patrolling in a float type seaplane."[4]

A further Memo noted that at a conference on Seaplane Schools on 31 December it was established that Calshot wanted no *Small Americas* but that Killingholme wished to increase its six *Small Americas* to 12.[5] The *Small America* boats were to serve out their lives training ham-fisted students in big boat flying.

Curtiss Model H-8

As noted, only one British boat was accepted as an H-8 as originally manufactured and delivered by the Curtiss Company. It is thought that this machine was delivered without a bow observer/gunner position, however this has not been confirmed, available photographs not showing the area clearly. All the remaining H-8 boats were delivered to H-12 standards.

A-152, the first H-12 ordered by the USN, was shipped to Pensacola on 7 March 1917, a year after the H-8 had been delivered to Felixstowe, yet this machine did not have a bow cockpit. All documentation refers to it as an H-12. Did this machine have an original H-8 hull? Again a lack of documentation and clear definition of aircraft types confuses the story.

Curtiss Model H-12

The Curtiss H-12 featured the laminated wood hull construction that was used for the *Small America* boats. The pilot and co-pilot sat side by side in the cockpit. If the need arose the co-pilot could crawl into the front cockpit to use the bombsight or the Lewis gun that was mounted on a pillar mount, and in later models, on a Scarff ring. The wireless operator sat facing forward behind the pilot on the right hand side of the boat. The W/T set could transmit for a distance of 80 to 100 miles depending on atmospheric conditions. When

Above: H-12B Convert N4339 as the rear windows define. The Gallows mounting appears to have multiple guns attached. The dark lower colour of the hull follows the line of the washboard. N4339 was apparently converted with an F.2A hull on arrival at Felixstowe in December 1917. Attacked by ten enemy seaplanes on 15 February 1918, the boat managed to flee without injury to the crew. On 1 July it was reported as worn out but was still at Felixstowe on 30 January 1919.

Curtiss H-12 Serial Allocation

Serials	Number	Contract	Contractor	Notes
8650–8699	50	CP77117/15	Curtiss, Toronto	Ordered as H-8 (2x160-hp Curtiss).W/E 02.04.17. Still listed as H-8 but with 2x250-hp Rolls Royce. Only 8650 delivered as H-8, rest H-12. Built in USA.
N1160–N1174	15	AS6731 (BR17)		Improved H-8. W/E 20.11.16 listed as H-12. Not delivered. Described as F.2 flying boat from 26.03.17 onwards. Presumed built as H-16, N4060-N4074.
N1505–N1519	15		AMC Ltd	Large America type. Cancelled.
N1510–N1529	20		AMC Ltd	Improved America type (2x310-hp Sunbeam) ordered January 1917. Apparent reduction of above. Cancelled by March.
N2330–N2353	24	AS6731? (BR106)		H-12. Renumbered N4330-N4353. (AS6731 apparently also covered BR50 for 60 and BR17 for 15 F.2A boats.)
N4330–N4353	24	AS6734 (BR106)		H-12B. N4351-N4353 cancelled.
Known H-12 & H-12B Converts: 8651, 8656, 8657, 8660, 8661, 8662, 8673, 8674, 8677, 8683, 8688, 8689, 8690, N4339, N4349.				

USN Curtiss H-12 Serial Allocation

Bureau No.	Number	Contract	Notes
A-152	1	SOC-490 Wash Requisition CR50; SE95; Bu-17	200-hp Curtiss V2-3. Crashed before accepted.
A-765 to A-783	19	SOC-625 CR119 SE249 Bu18	330-hp Liberty.

Specifications H-8/H-12

Type	H-8[1]	H-8[2]	H-8[3]	H-12[4]	H-12[5]	H-12[6]	H-12[7]	H-12[8]	H-12[9]	H-12B[10]
Span Top	91 ft 0 1/8 in		92 ft 10 in	92 ft 8 9/16 in	92 ft 8 9/16 in	93 ft 0 in	95 ft 2 in		96 ft 6 5/8 in	95 ft 0 13/16 in
Span Bottom			66 ft 11 in	66 ft 11 3/8 in	66 ft 11 3/8 in				68 ft 11 3/8 in	68 ft 11 3/8 in
Chord	7 ft 0½ in				7 ft 0 9/16 in				7 ft 0 49/64 in	7 ft 0 49/64 in
Gap					8 ft 0 9/16 in				8 ft 0 9/16 in	8 ft 0 9/16 in
Incidence					4°				4°	4°
Dihedral	1°				1°				1°	1°
Section					RAF-6					RAF-6
Length Overall	40 ft 6 in/ 42 ft 6 in**		44 ft 8 in	46 ft 5½ in	46 ft 5½ in	46 ft 0 in	46 ft 6 in		46 ft 5½ in	46 ft 5½ in
Height						16 ft 9 in	16 ft 6 in		16 ft 9 5/16 in	16 ft 9 5/16 in
Length Hull					42 ft 0 in					
Width hull Over Fins					10 ft 0 in				10 ft 0 in	
Hull Depth					6 ft 8½ in				6 ft 8½ in	
Areas, Square Feet										
Upper Wing		578							506.20	
Lower Wing		394							432.60	
Wing Centre Section Top		70							95.13	
Wing Total							1,216	1,108		1,164
Ailerons		92							133.00	131
Horizontal Stabiliser		78							108.00	108
Elevators		56							58.40	58.4
Non-Skid Fins									26.60	26.6
Fin		26							30.80	31.1
Rudder		35							32.40	32.4

Specifications H-8/H-12

Type	H-8[1]	H-8[2]	H-8[3]	H-12[4]	H-12[5]	H-12[6]	H-12[7]	H-12[8]	H-12[9]	H-12B[10]
Weights in Pounds										
Empty	5,200						7,293	6,351		6,500
Fuel & Oil							2,057	1,734		1,635
Crew							720	540		360*
Armament							580	467		0
Useful Load										2,900
Gross	7,500						10,650	9,092		9,400
Capacity in US Gallons										
Fuel										249.5
Oil										10
Speed										
At 2500 ft							87 kts			
At 10000 ft							81 kts			
Maximum	72 mph							78 kts at sea level		95 mph
Endurance										5.1 hrs
Climb in 10 mins										4,200 ft
Time to 2,000 ft							3.3	4:00		
Time to 6,500 ft							13.7	16:00		
Time to 10,000 ft							29.8	36:00		
Service Ceiling in ft							10,800			

* Crew of two. Usually more were carried. ** This dimension is hard to decipher on the document and may be incorrect.

Notes on Sources for H-8/H-12 Specifications:

(1) Curtiss Specification sheet for H-8 tractor dated 08.04.16. NASM File AC-901660-01 Curtiss Model H-8 Small America.
(2) "Design of H 8 T.B. – Experimental, Report on Preliminary Layout." From Chance M. Vought. This data sheet has "Tractor Type Admiralty requires" on page 2, where weights are estimated. Casey states that this is the corrected carbon copy. Copy in RAFM J.M. Bruce Collection Box 9.
(3) RAFM J.M. Bruce Collection Box 22. Felixstowe File. Typewritten specification chart.
(4) *Ibid.*
(5) Dimensions recorded by C. Roloson, Pensacola, 23 March 1917. Aircraft Collection Special Section, Box 294. NHHC, Washington Navy Yard, DC.
(6) *Types of British Seaplanes*, copy in TNA. The boat illustrated is 8681. The use of a wider centre section would account for the different wing spans.
(7) *Record of Performances of British Seaplanes 1917–Dec'31 1918.* Copy in TNA AIR1/708/27/11/02.
(8) Air Board "Table of Performances of Seaplanes & Flying Boats, 31 Oct 1917". Copy in RAFM J.M. Bruce Collection Box 26.
(9) Specification Sheet for H-12B. "Curtiss flying boat, enclosed cabin, lines generally identical with H-12A hull." Copy in NARA.
(10) USN Chart: *Characteristics of Flying Boats and Seaplanes*, May 1918. Copy in NHHC.
Dimensions are as stated in original documents.
A report stated that the H-12 *Large America* carried:

- A crew of four.
- Usually 4, sometimes 6 Lewis guns.
- 2 x 230-lb bombs.
- Wireless.

Range in a calm was approx. 400 sea miles.[12]

Above: According to the serial on the beaching trolley this is Curtiss H-12 8682. If it is indeed 8682 then it has been upgraded to an H-12 Convert with an F.2A hull. Note the discarded H-12 hull in the background.

in the air a weighted copper wire aerial was lowered through a tube that ran through the hull. This could not be used on the water, so a telescopic wooden mast about 30 feet high was raised. An aerial was led from the bow, tail, and ends of the upper wing to the tip of the mast. In this way a message could be sent for a distance of about 30 miles. The wireless operator also coded and de-coded all messages. The codebook was bound with lead in the covers so that it could be dropped over the side in the event of capture. An Aldis lamp was available for signalling with ships and other flying boats, emergency rations, the first aid kit and the cage for the carrier pigeons were all the province of the W/T operator.[6] The fourth member of the crew was the engineer who looked after the engines, fuel and occupied the rear gun layer's position when required.

The wings followed the same plan as the earlier boats but were increased in size. The wings of the *America* boats did not undergo a similar development process as that applied to the hulls and the construction remained the same with the only size changing. Late modifications to the F.2A and F.3 included balanced ailerons.

Pix Hallam recalled that the War Flight had a boat turned over to it for the purpose of instructing pilots on the big boats. In their "spare time" 1st pilots were expected to train 2nd pilots "in the gentle art of taking off and landing a big boat –

Above: USN H-12 boats were Navy Grey overall as shown here on A-782. Reported as 100% complete in April 1918 A-782 was shipped to Hampton Roads from 11th to 21st of that month. It was delivered with material shortages that kept it OOC, a not uncommon occurrence with the USN contractors during the war. Hampton Roads reported the machine as commissioned on 8 June. It performed an anti-submarine patrol under Lt Townsend on the 14th. These continued through to July when on 24th it had accompanied A-783 on patrol when an oil leak occurred when 60 miles out to sea. By throttling down the motor they were able to complete the patrol and alight safely inside "the Capes." After the flight, oil seen to be leaking from the cam-shaft housing. On glide downward, engines idle, the propeller and pair of connecting rods came adrift. Four bolts broken, bronze bearings broken in small pieces, one cylinder & both connecting rods deformed, both pistons shattered, upper and lower crank case broken. "Spare parts should be furnished to rebuild this engine." On 3 September a forced alighting in a rough sea was caused by the starboard engine failing in flight. The vertical struts inside the hull were bent and broken; the front spar of the right wing was cracked inside of the hull; ribs and the forward compression strut from hull to right wing was bent. Hampton Roads recommend strengthening of H-12 Hulls. It was shipped to the NAF rather than Curtiss for repair. Despite this the machine was back on anti-submarine patrols on the 18th. On 21 September 1919, Hampton Roads reported that the machine was with Squadron 1 and A-783 had, with Machinist Skyles and three men, attained 10,600 feet. "Has been in commission since 8th June and oldest boat on East Pier." The boat was stricken on 20 August 1919.

Below: The Bureau No. of this USN H-12 is not carried on the hull. The H-12 was mostly used for training in the US.

Above: With an original caption stating this is also the F.1, this Felixstowe boat has a revised rudder and forward gun ring.

an exercise which proved equally hard on the nerves of the instructors and on the bottom of the machine, as there was only one control wheel fitted and the first pilot had to give up all control to the pupil."[7]

The first H-12 boats had a narrow front gunner's cockpit of oval shape and the gunner had a spigot mount for his Lewis Gun. The main cockpit was enclosed almost completely by celluloid panels that had an opening on the port side for entry. The hull had a more sharply pointed bow from the H-12 boats that followed. The later boats had a circular front cockpit with gun ring and the bow was not as pointed. The pilot's cockpit was now covered by a canopy of flat panels that was separated from the main body of the hull by a prominent gap. The H-12B apparently had a wider centre section. Sir Austin Robinson thought that the gap between the engines was increased from 11 ft 9 in to 13 ft to facilitate the installation of 310-hp Sunbeam engines, although no evidence of such an installation to an H-12 has been found to date.[8]

The H-12 specifications are one of the most confusing of the *America* boats as will be seen in the following tabulation from primary sources.

On 14 March 1917, Hope Vere from Tresco Station in the Isles of Scilly, sent a report on the H-12 where he expressed concern that although the "machines are standing up moderately well" they were having "a great deal of trouble with leaky tails."

The trouble appears to be that lying in the water all the time softens the wood to such an extent that wherever screws are used instead of rivets in fixing planking to floors the screws pull out when getting off the water due to suction on the tail. I am replacing all screws with rivets and hope this will be a great improvement.[9]

Although he had not lost any machines "we have had two very near squeaks." 8656 was reported as sinking one night and by working all night and the next day the H-12 was beached. Repaired, the machine was ready for work again. 8652 took off from Newlyn for Brest and her tail planking pulled off due to the above reasons. This H-12 was beached and had to be disassembled there. Hope-Vere was presently waiting for a trailer to transport the seaplane back to base. 8653 had not arrived as it had broken down and was at Calshot being repaired.

"The planes are standing up to the weather wonderfully" as were the engines that were "as a rule," easier to start than when "they lived in a warm shed." Considering the weather and rain that they experienced nearly every day, he was of the opinion that the "the experiment bids fair to be reasonably successful." It is considered that he was referring to the operation of these large flying boats without a large base with hardstand and hangars, instead they were moored out in the weather. Hope-Vere was anxious to get Felixstowe hulls instead of the *America* hulls. "With Felixstowe hulls & a well protected anchorage I see no reason why seaplanes of the boat type should not be moored out for an indefinite period." He finished his report by stating that if he could get Felixstowe hulls "I can guarantee some useful work."[10]

Above: Felixstowe F.1 3580 after its conversion from a Curtiss H-4 using Porte's hull design. It was at the Felixstowe Seaplane School by 29 December 1917, not being written off until early 1919. Note the pitot tube extending from the bow.

Felixstowe F.1

The Felixstowe F.1 was the name given to John Porte's experimental flying boat that had an experimental Porte designed hull and H-4 aerostructure. The hull was built on aircraft lines with cross braced girder of simple longerons and spacers to which was attached the bottom of the hull. Power was provided by two 150-hp Hispano-Suiza engines.

Unfortunately on test the following result occurred. "On practically smooth water and with no wind, both with the elevator full up and down, the flying boat failed to unstick the tail."[11]

Specifications

Length hull 36 ft 2 in. Wing Span: Upper 72 ft. Lower 46 ft. Chord 7 ft; Gap 7 ft 6 in. Engines: Two 150-hp Hispano-Suiza.

Felixstowe F.2

The evolution of the F.1 and F.2 is described in Chapter 2. 8650 was reported as undergoing overhaul between 14 August and 4 September 1916, and it is likely this is the period it was fitted with the Porte hull. The combination of a new Porte hull with the H-12 wings and a new tail unit led to a better boat that was designated F.2 although it retained the serial 8650. With more powerful engines it was developed as the F.2A.

The F.2A Described

The hull was composed of four longerons with horizontal and vertical spacers braced by diagonals and steel cables. The planning bottom was constructed of two-ply cedar inside and mahogany outside. Varnished or doped fabric was sandwiched between the two layers. The bow, decking and cockpit framework was mahogany while the sides back to the trailing edge of the lower wings was birch plywood. The top of the hull behind the wings was fabric covered to save weight. A 12 inch deep mahogany washboard ran along the lower longerons. The front part of the hull as far back as the rear spar was a rigid built-up structure. From here to the rear the structure was a typical fuselage structure with doped fabric covering above the 12 inch solid mahogany washboard. This was to cause problems, and following the loss of H-16 N4510 on 8 April 1918, in heavy seas, due to the failure of the fabric sides on the tail of the boat, the USN recommended that this defect be corrected on all its H-16 in service.[12] The fabric was to be replaced on some boats with ply sheeting. The fins were built onto the outside of the hull and were planked on the top with three-ply. The hull was double planked. Most WWI built aircraft show variations however the Felixstowe boats show more than others. This was due in part to the different manufacturers of hulls and the aerostructure being allowed considerable latitude and the modifications worked out in

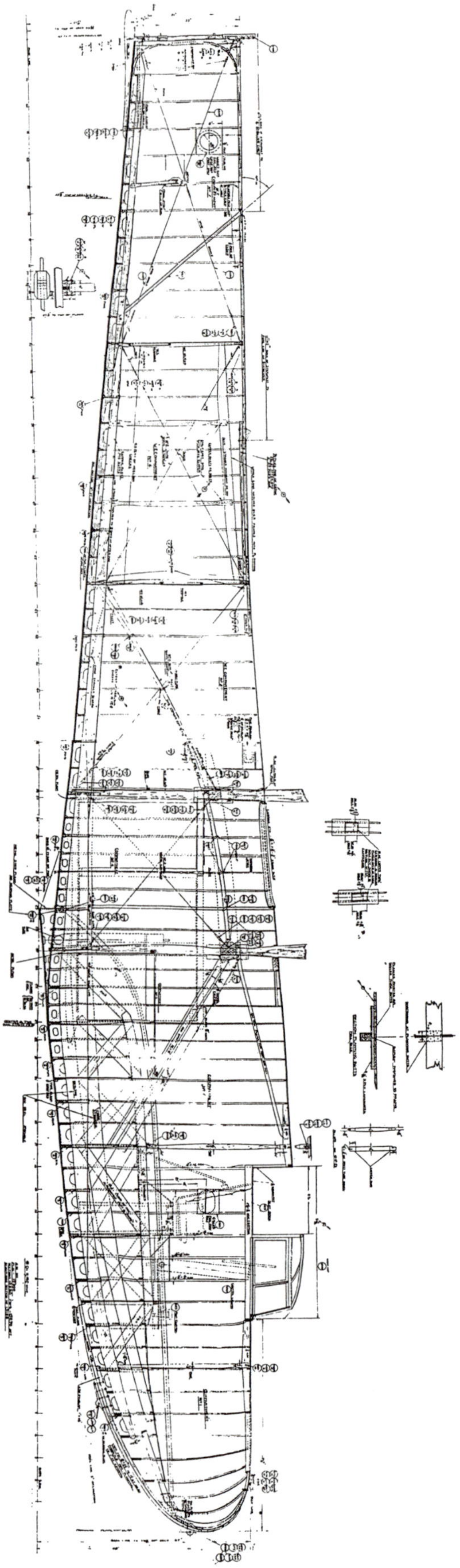

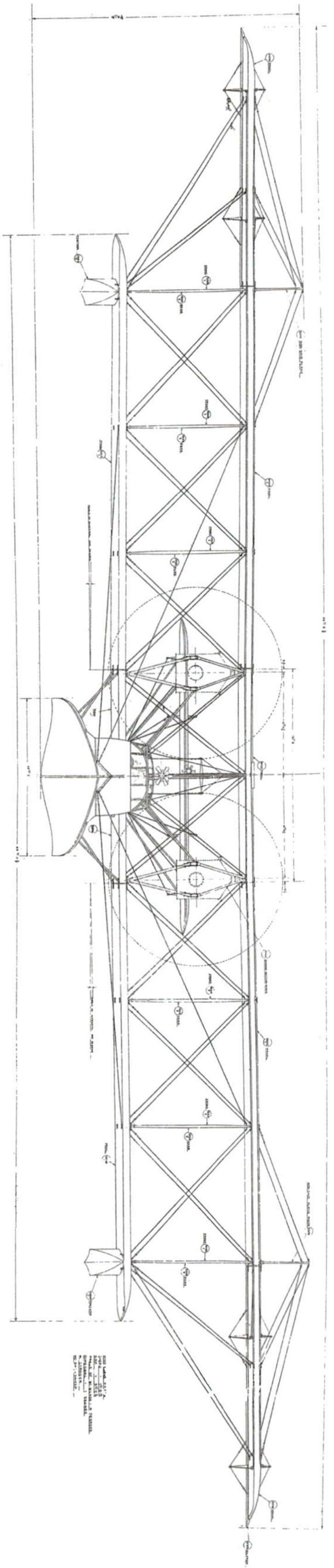

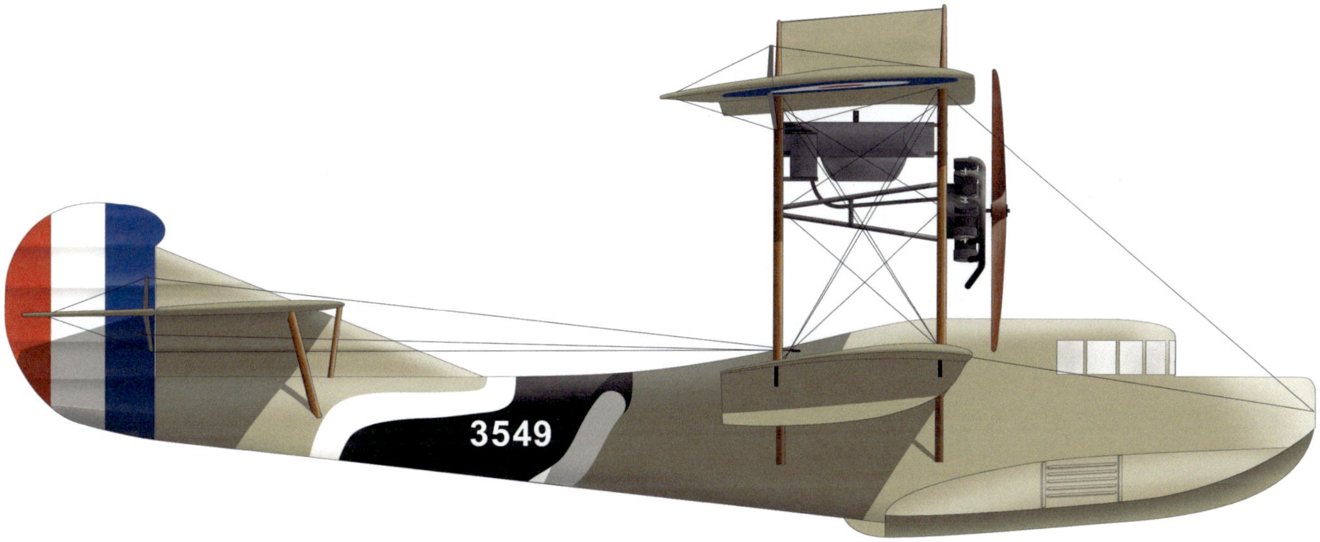

Curtiss H-4 3549 was one of the order for 50 Small America boats ordered in March 1915. It was due for deletion in December 1917 but survived at the Felixstowe Seaplane School well into 1918.

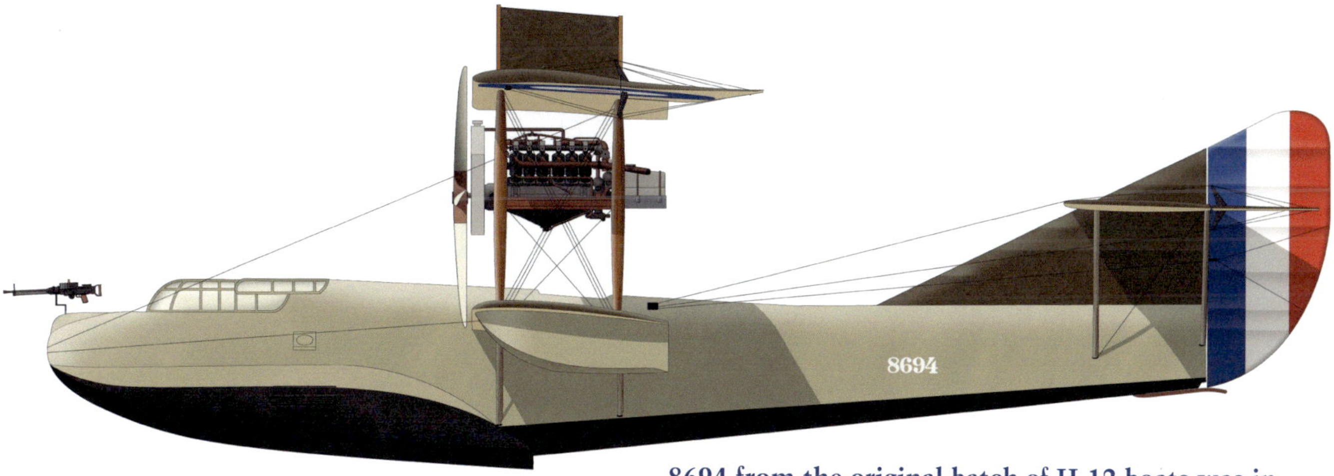

8694 from the original batch of H-12 boats was in service from at least September 1917 when it was engaged in an attack on a submarine. Damaged beyond repair it was deleted on W/E 17 November 1917.

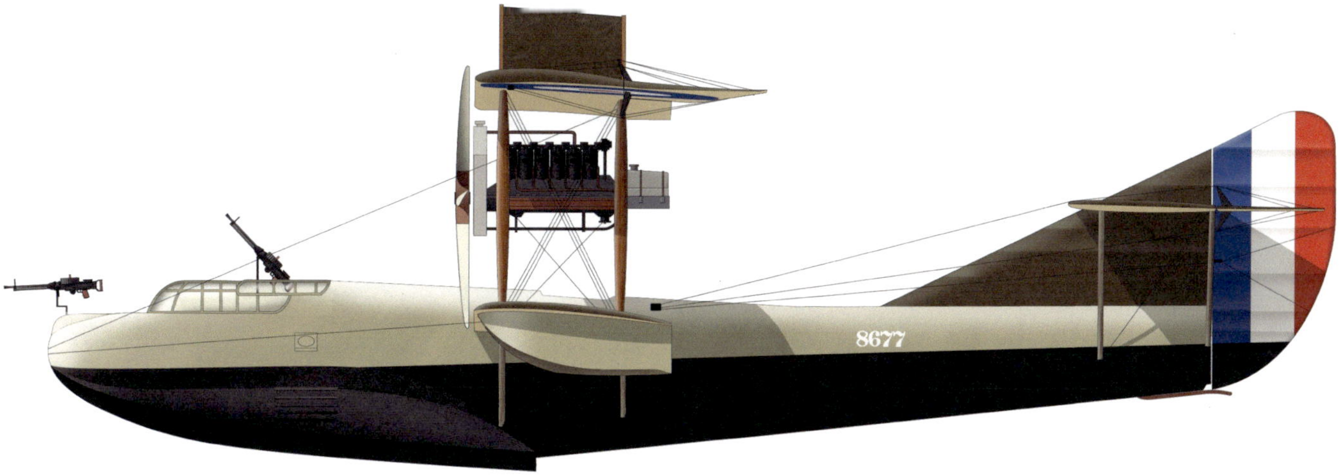

Curtiss H-12 8677 assigned to Felixstowe.

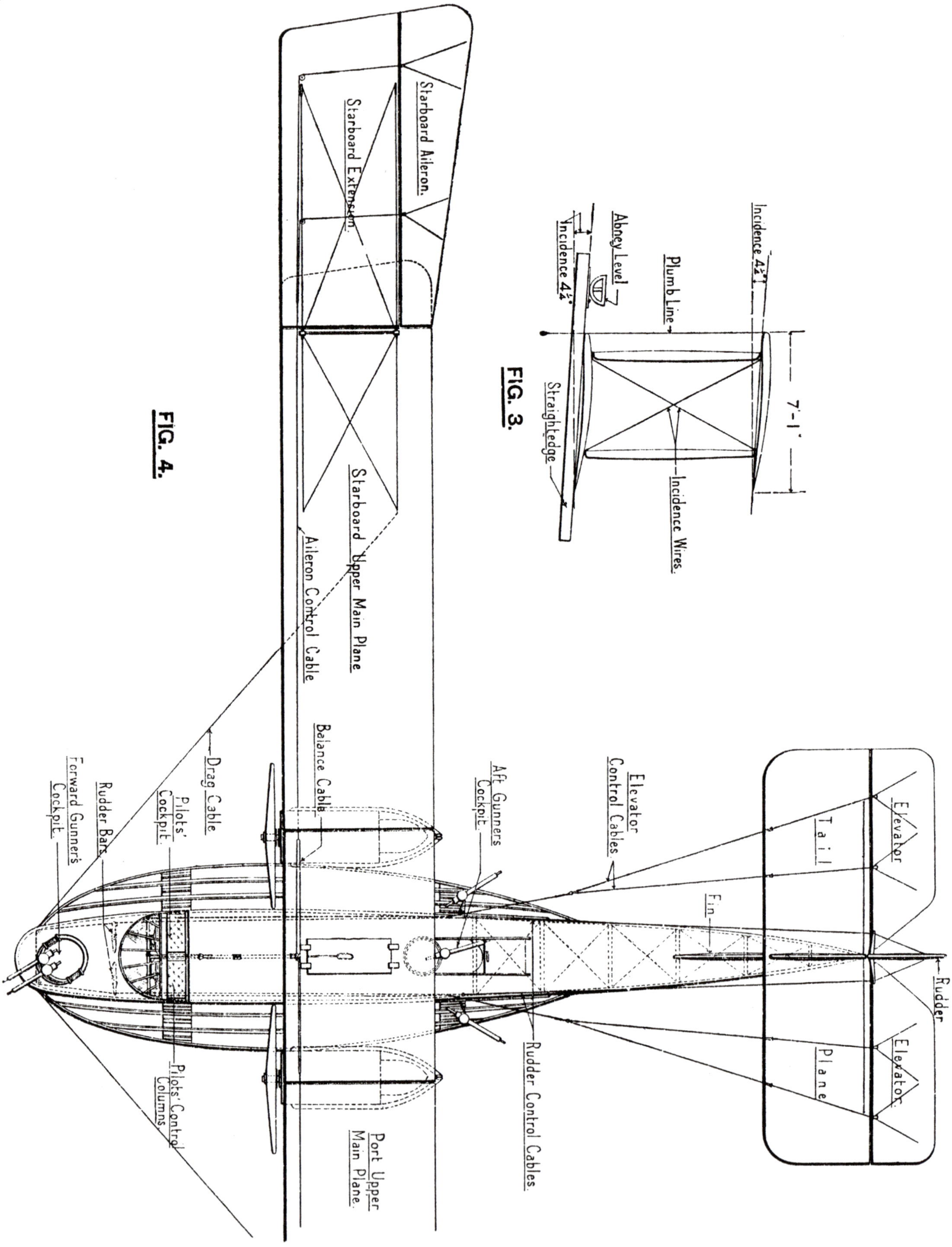
Incidence 4½°
Plumb Line
Abney Level
Incidence 4½°
Straightedge
7'-1"
Incidence Wires.
Starboard Aileron.
Starboard Extension.
Starboard Upper Main Plane
Aileron Control Cable
Balance Cable
Drag Cable
Pilots' Cockpit
Rudder Bars
Forward Gunner's Cockpit
Pilots' Control Columns
Port Upper Main Plane.
Aft Gunners Cockpit
Elevator Control Cables
Rudder Control Cables.
Fin
Tail
Plane
Elevator
Elevator
Rudder

FIG. 3.

FIG. 4.

Felixstowe F.2A

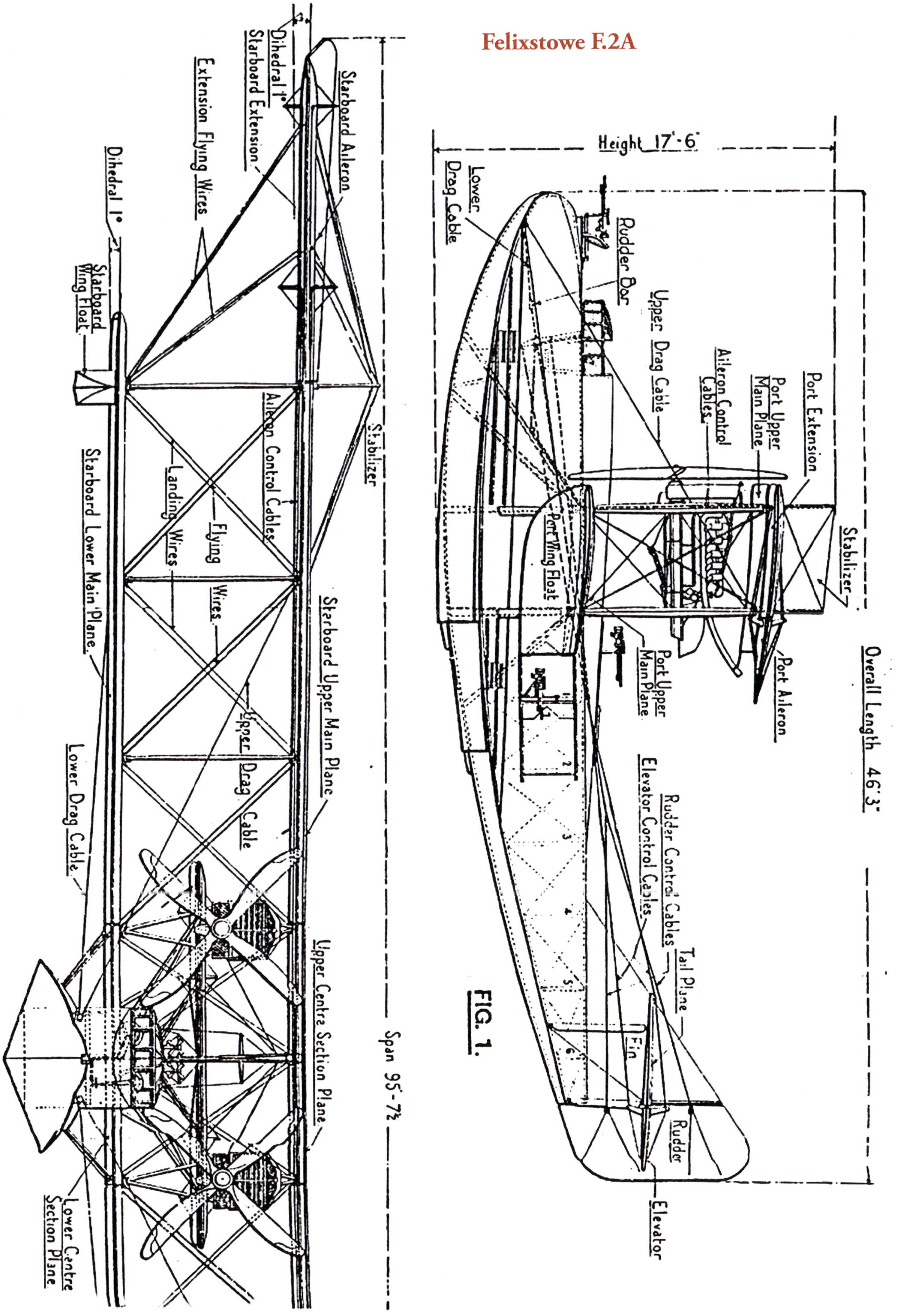

Above: Close up of the Gallows mount Lewis gun deployed. The crewman has to adopt this position to gain full coverage from this mount. By cocking up the hull aft of the wings Porte gave the gun layers a better field of fire than the armament of the Curtiss H-12, however there were still areas where an enemy could hide and spray the flying boat with fire. The boats would fly close to the water when attacked to prevent these attacks from behind and below.

active service and applied at stations.

The hull was V-shaped and curved from bow to stern while the tail portion was raised such that it rose clear of the water when at rest. The side fins were built onto the outside of the hull. The planking was double diagonal comprising an inner layer of ½ inch cedar and an outer layer of $^{3}/_{16}$ inch mahogany, the two with a layer of varnished fabric sandwiched between them.[13] Sliding panels in the hull behind the wings allowed for a Gallows mounting for a Lewis Gun on each side. These could be swung outboard and covered the tail of the boat thus overcoming one of the H-12 defects. A pilot of the F.2A could alight in much rougher seas with less fear that the hull would be damaged, and he could take-off in rougher waters as it did not hammer the water to anything like the extent that the practically flat bottom of an H-12 did. This was despite the F.2A being designed by Porte to utilise the sheltered waters of harbours such as Harwich, the necessities of war calling for more from the boat than its designer intended.

The wings were a typical Curtiss wooden structure and plan form although the F.2A boats had ailerons that projected beyond the wing trailing edge. Late boats had open cockpits and balanced ailerons.

The decision to change from a 23 inch to 20 inch gun ring involved structural alterations and delayed production, the type entering service late in 1917. The number of *Large America* boats required for the 1918 program was estimated to be 180 in May 1917, and this was revised to 426 with the revised program of July 1917. The average life of the big boats was six months and therefore twice the number of boats was required to meet the estimated establishment figure. The entry of the US into the war and the USN's agreement to take over and equip five naval air stations provided some relief to the situation as even with the Porte method of construction, it would have been impossible to produce the number of boats required. In March 1918, 161 F.2A and F.3 boats were on order, however only ten F.2A and one F.3 were in service.

The F.2A suffered with problems with its fuel system. The length of piping it contained ran from the main tanks in the hull where wind-driven pumps forced the fuel to the gravity tanks in the top wing centre section. From here the fuel was fed to the carburettors of the Eagle engines.

From about September 1918 all the new F.2A had open cockpits, the canopy being done away with. They were slightly faster as a consequence and the pilots had a better view backwards. The fabric was deleted from the rear of the hull and replaced with Consulta sewn ply. Horn balanced ailerons were introduced making them less tiring to fly.[14] The F.2A had dual control unlike the H-12. The co-pilot's control wheel folded enabling him to leave his position and access the front gun cockpit.

Above & Below: An F.2A in flight. The gunner in the rear cockpit at the wing trailing edge can be seen.

Below: This F.2A broke in two and turned turtle. The hulls of the F.2A varied in strength according to manufacturer and the naval air stations would modify them to suit their needs.

Above: Twin guns on the nose cockpit of this F.2A.

Below: An F.2A in trouble.

Above: A portrait of the pilots in flight. **Below:** A view from the rear cockpit. Note that the hull is painted in angled stripes.

Above: N4446 on its handling trolley with that of 4013 in the foreground. Note the position of the serial on the rudder. A Saunders built boat, N4446 Was not delivered until after January 1919.

Below: N4465 with service and civilian personnel. There is a stencilled line of small text under the serial indicating a Saunders built hull. N4465 was in the same batch as N4446 but has the serial on the top of the rudder. This machine was at Killingholme in 1918 with balanced ailerons and no cabin. Nothing further is known about its career.

Above: Pristine F.2A with cockades under the top wing.

Below: An F.2A gets up onto the step.

Above: N4453 under construction at S.E. Saunders. There is no record of this machine being delivered and it may have gone directly into store. Note the lack of a cockpit canopy.

Below: An F.2A hull in the Aircraft Manufacturing Co works. Note the stencilling on the base of the fin.

Above: Beached F.2A with cabin enclosure.

Below: Three beached F.2A flying boats are visible in this photograph.

Above: All hands on deck! The number of personnel to operate these large seaplanes is illustrated by this photograph of N4291 *OLD BLACKEYE*. This machine has the lower cockade painted under the lower wing.

Below: N4436 on the slipway. This machine was recorded with No.267 Squadron at Calafrana. The white forward hull deck fabric was probably adopted as it let light through to illuminate this section of the hull where the radio operator, etc., were located, and at the same time protected the fabric from the rays of the sun. V.84 aluminium would have been opaque and not provided this benefit. (Nicolas Cooper via SEAWINGS)

Above: An F.2A on the slipway.

Above: Dazzle painted N4541 and N4504 (?) trail this line-up of F.2A boats on the Felixstowe slipway. Note that N4541 retains its dark coloured hull top decking.

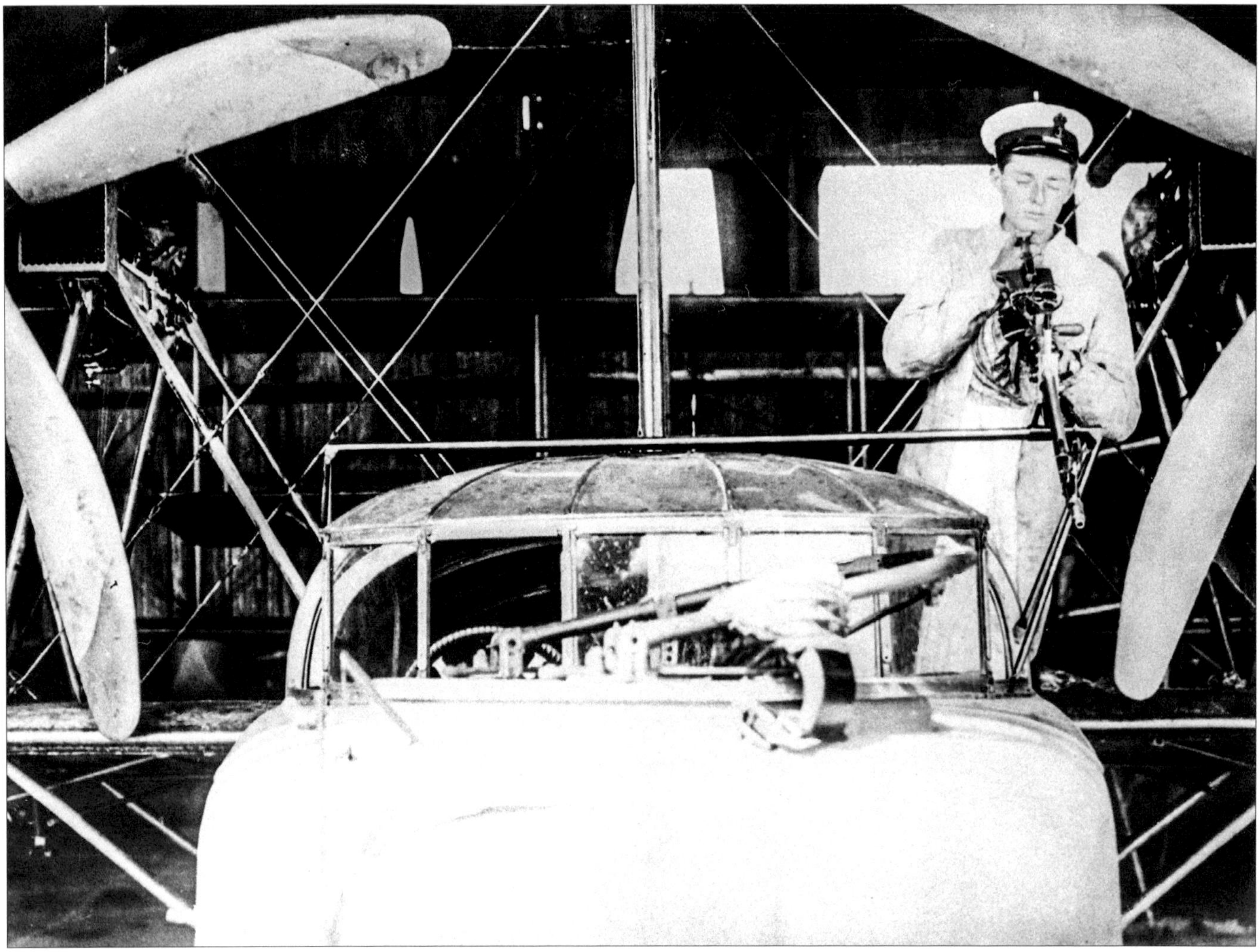

Above: An experimental or squadron mount to allow the second pilot a machine gun. One can only imagine the front gunner's comments to this bizarre arrangement. (Nicolas Cooper via SEAWINGS)

Above: N4430 has its forward hull dazzle painted as are the wing tip floats. Note the low serial application to the rudder.

F.2A Specifications

Source	1	2	3	4	5	6
Span Upper	95 ft 7½ in	95 ft 7½ in				95 ft 8 in
Span Lower	68 ft 5 in	68 ft 5 in				
Chord	7 ft 1 in	7 ft 1 in				
Incidence	4½°	4½°				
Dihedral	1°	1°				
Length	26 ft 3 in (sic)	46 ft 3 in				46 ft 3 in
Height	17 ft 6 in	17 ft 6 in				17 ft 2 in
Wing Area					1,133 sq. ft.	
Weights in Pounds						
Empty					7,549	
Crew 4			720		720	
5 Lewis Guns, mounts, ammunition			293			
Bomb gear, Aldis lamp, sea anchor, etc.			292			
Total military load			1,305		585	
Fuel & Oil					2,124	
Gross			11,884	9,498	10,978	
Climb Time to						
6,500 ft			16 min 30 sec	10 min 30 sec	16.7	
10,000 ft			39 min		39.5	
12,000 ft				23 min 30 sec		
Speed						
at 2,000 ft					83 kts	
at 6,500 ft					77 kts	
at 10,000 ft					70 kts	
Endurance					6 hours	
Ceiling					6,500 ft	

Sources:
1. Aeronautics 29.01.1919.
2. F.2A Rigging Notes.
3. Summary of Reports in USN Technical Note 57. Report N.M. 100. (N4510).
4. Summary of Reports in USN Technical Note 57. Report N.M.104. (N4510 – without bombs).
5. *Flight* 22.04.1920.
6. *Record of Performances of British Seaplanes 1917-Dec'31 1918*. Copy in TNA AIR1/708/27/11/02. (Trial report March 1918)

Engines: Two 345-hp Rolls-Royce

F.2A Serial Allocation					
Serials	**Number**	**Contract**	**Manufacturer**	**First Delivery**	**Notes**
N1260–N1274	15	AS3610 (BR17*)	Curtiss, Canada.		H-16. Presumed renumbered N4060-N4074.
N2280–N2304	30	AS14154 (BR80)	S.E. Saunders Ltd.		Renumbered N4280-N4309.
N2530–N2554	25	AS21558	AMC Ltd.		Renumbered N4530-N4554.
N4080–N4099	20	AS14154	S.E. Saunders Ltd.		Delivered from late July 1918.
N4280–N4309	30	AS14154 & AS34426 (BR80)	S.E. Saunders Ltd.		Delivered from mid-November 1917.
N4430–N4479	50	AS4498/18 (BR349)	S.E. Saunders Ltd.	18.10.18	Some delivered as F.5s.
N4480–N4504	25	AS4502/18 (BR350)	May, Harding & May	21.09.18	Erected by AMC Ltd. At least 21 delivered.
N4510–N4519	10	AS2697	May, Harding & May	12.01.18	Specification N.3B. Sunbeam engines contemplated for this batch.
N4520–N4529	10	AS24912 (BR74 & BR90)	May, Harding & May		Cancelled.
N4530–N2554	25	AS21558	May, Harding & May		Erected by AMC Ltd?
N4555–N4559	5	AS24912			Cancelled.
N4560–N4579	20	38a/551/C564 & AS24912/18 (BR589)	May, Harding & May		Last seven cancelled.
The cost of an F.2A including hull and trolley, but without engine and armament was £6,738. Eagle VIII engines cost £1,622.10.00 each.					

Above: F.2A N4510 on its beaching dolly.

F..A Flying Boat N.4383 – Getting Off Tests on 09.05.1918

Object of the Test

The object of the test was to find the condition of taking off (1) with varying horsepower and the same load (2) with maximum horsepower and varying load. Some trails were made outside the harbour with a moderate swell, but those reported were all done inside the harbour with smooth sea and a 7.8 knot wind. It appears as if the taking off on a choppy sea is to a high degree fortuitous and there is more likehood of obtaining results which can be interpreted in the case of smooth water.

Apparatus and Method

A Badin ASI was provided with a powerful venturi to record airspeed. The clock was started on the instant of opening up the engines. The time of getting off was obtained by stop watch by an observer in the rear cockpit watching the tail and an observer by the side of the pilot by a sign from the latter.

The load was varied by changing the amount of fuel and taking on lead and passengers. The max load of 12,450 lbs was made up with full tanks, 12 men and 400 lbs of deadweight. This was considered the limit on the margin of safety for these tests.

In the case of varying horsepower, the throttle was opened up from dead closed to different degrees and kept constant until the machine was well off the water. The RPM was measured before hydroplaning.

Results

There is a very distinct change of acceleration at one point that appears to be the point at which the boat gets on the step and begins hydroplaning. The actual takeoff is not shown in any characteristic way. Speed is increasing uniformly or constant when the machine takes off.

Varying RPM – Load 9,235 lb

RPM	Time to Get on Step in Seconds	Time to Get Off in Seconds	Accel in Taxiing Kts/Sec.
1300	Failed to get off		0.18
1325	28	58	0.72
1370	25	38	1.16
1440	20.5	23	1.00
1510	15	21	1.00
1540	14	18	1.54
1615	11	15	1.80

Varying Load – Engine Full Out

Total Wt in lbs.	Lbs/H.P.	Lbs/sq ft	Time to Get on Step in Secs.	Time to Get Off in Secs.	Accel in Taxiing Kts/Sec.
9,668	14.00	8.54	15	24	1.56
10,025	14.55	8.84	16.5	25	1.56
10,383	15.03	9.14	17	27	1.54
10,980	15.90	9.67	18	20	1.68
11,429	16.58	10.09	21	24	1.14
11,685	16.91	10.3	-	23	1.36
11,940	17.31	10.51	19.5	27	1.42
12,197	17.69	10.76	17	37	-
12,450	18.06	10.98	20	30	0.75

Source: Secret Technical Information, 15th May 1918. Copy in RAFM J.M. Bruce Collection Box 64.

Above: N4892 on the beach. This Curtiss H-16 appears to have had the ply painted the same dark colour as the fabric surfaces rather than being varnished British fashion, this dark colour continuing onto the hull bottom. Lack of documentation makes it difficult to make reliable assumptions as to the colours of these flying boats. Delivered in September 1918, N4892 was from a batch of 110 boats ordered from the USA but only N4890 to N4949 were received, the rest being cancelled.

Curtiss H-16

As noted before the Curtiss H-16 was the Felixstowe F.2A built in the USA with Liberty engines. The H-16 was naturally selected to be the type that the USN would take into action from the stations that they were to establish in the UK. The Curtiss Co had a contract with the Admiralty for 75 H-16 boats without engines and with the USN for 34 with Liberty engines. The Bureau was "acting on the principle that we should interfere as little as possible with the British production, which should have priority." The estimated production of the Austin Street plant was three per week. It was proposed that delivery between the Navy and Admiralty that the proportion of assignments of H-16 boats be one to two. As things turned out, the British did not utilise their H-16 boats as much as they intended as the first one delivered, N4060, was 500 lb. overweight and Porte's examination concluded that they had increased weight little by little all along. The flying boat was found to be tail heavy on arrival and the after petrol tank on the port side was moved forward. "In this condition the machine is in balance." N4060 was described as an anti-submarine and long reconnaissance machine in Report NM200. It underwent testing at Grain in May 1918. The observer had an "excellent view forward and the sighting and bomb releasing positions are good." The cabin was provided with celluloid windows instead of triplex glass. The pilot's view was so bad that the cabin was removed and replaced by an ordinary wind screen.

The machine was slow to manoeuvre and the ailerons, as well as the rudder, were very heavy making the machine tiring to fly. Under the heading of directional stability it was stated that the rudder must be used with lateral control otherwise it caused "pawing." The boat was easy to alight but the nose tended to rise out too much as the planning speed fell. The machine had to be modified before it could be considered ready for active service. The step required alteration, the fuel tanks had to be moved forward to overcome the tail heaviness, dual controls and the Second Pilot's seat had to be installed, and balanced ailerons fitted.

The Report on N4060 noted that there were four fuel tanks behind the pilot of 62 gallons each. They were operated on a pressure system using wind driven pumps to force the fuel to the 225 gallon gravity tank in the centre section. A 12 gallon oil tank was situated behind each engine. It took 15 minutes to get under way. Ease of starting was described as "good."

A Report in August noted that "F.2a Boat H-16 No. 4,060 (two Eagle VIII Rolls-Royce)" was presently having balanced ailerons, elevators and rudder fitted and was to undertake investigations of the slipstream near the tail. The same Report noted that N4892 was to undertake comparative trials with N4060. A Wright's vertical fin balance was being fitted and "will do investigation of pressures of hull."[15]

N4892 was tested with the Wright stabiliser. This device comprised two vertical surfaces hinged at their after edges on to one of the front struts of the flying boat. They were connected

Above: Curtiss Aeroplane & Motor Corp H-16 in the Curtiss plant at Buffalo, NY. According to Sir A. Robinson, the aluminium step treads on the walkways over the fins were a distinguishing feature of the H-16 on the ground. This is most probably N1162.
Below: N1162 in the same location. Note the flap in front of the aileron gap that closed with air pressure in flight. N1162 was not delivered as such. It is thought that the batch N1160–N1174 was delivered as N4060–N4074.

Above & Below: N4060 without a cockpit canopy. This was the first of a batch of 15 H-16 boats from Curtiss. It arrived at Felixstowe in March 1918 and was at the Isle of Grain for testing and experimentation from May. It then served with No.230 Squadron, RAF, and disappears from the records in January 1919.

Above: N4072 shows the fabric covered rear fuselage above the wash board well in this colour scheme. The rudder is unbalanced and a cockpit canopy is fitted. Note the walkway on the lower wing outboard of the engine support struts. These machines have a dark colour to the upper fabric surfaces of the hull and wings. The Colour that Curtiss used is unknown at this time but may have been the green that is stated to have been applied to the *Large America* boats.

Below: Dazzle painted British H-16 with balanced rudder and ailerons and cockpit canopy removed. Note what appears to be a large weapon in the rear cockpit. Another dazzle painted boat rides at rest further out in the water.

Below & Bottom: The first H-16 from the NAF is ready for launching. The early H-16 boats had a gooseneck spigot mount for the front Lewis gun. This was replaced by a Scarff ring on the boats that went to Europe.

Above: H-16 hulls in the NAF. This photograph gives some idea of the size of the USN's production program.

to the ailerons so that the movement of the ailerons displaced the vertical fins. The fins were to act as a balance device for the ailerons, and in addition any sideslip would put a load on the fins and cause them to move the ailerons so as to restore lateral balance.

The device was flown on 21 November 1918, and found to be very sensitive. "As it seemed that the machine might take charge if any gusts were encountered, the pilot landed after a short flight." The surfaces were reduced to a half the original area and tried again on the 30th by Capt Moore, and on 2 December by Maj Cooper. "The machine was flown for about quarter of an hour without the hands on the wheel and turns were made, more or less sideslipping taking place. The behaviour of the machine was unpleasant to feel and in gusty weather the device would be unstable." It was considered that a machine such as the H-16 was not suited to the device "as the rudder has to be used a great deal in conjunction with the ailerons."[16]

Britain ordered 125 H-16 boats from Curtiss; the sometimes quoted 185 figure includes 60 that were reserialled. On 31 October 1918, the RAF had 32 F.2A boats, 6 H-12 Converts and 69 H-16 boats on charge, all in the UK; none were at bases overseas. Of these 31 of the H-16 boats were in store.

USN H-16

In July 1917 the Bureau of Construction and Repair of the USN (Bu C&R) was notified by their superintendent Constructor of Aircraft at the Curtiss Aeroplane Co, Buffalo, NY, that Curtiss were constructing a new flying boat, the H-16. The letter then goes on to describe the Porte method of

45
845

Above & Facing Page: H-16 A-845 with aluminium doped wings and tail, and post-war insignia. The "45" is marked on the upper centre section and on the bow. Note the modified balanced rudder. Reported as 25% complete in June 1918, and 70% complete in August, A-845 was 100% complete in October and was assigned to Pensacola. It was assessed as too bulky to rail and was assembled at Hampton Roads and flown to Florida. Pensacola reported on 19 May 1921, that the boat was in storage and would be converted to dual when erected. This was to be done as there were not enough F-5-L boats on the station and although the policy was to replace the H-16 with the F-5-L, the H-16 boats were required to complete the station complement for training. A-8445 was in service by 15 December 1921 at least. It was in continuous service training until 10 June 1922 when it was transferred to the NAF by air, arriving on the 11 July. On 4 June 1923, the NAF reported that the machine had been in storage since manufacture for about 5 years, which was incorrect. The hull and surfaces showed signs of deterioration due to age, etc., and it was recommended that it be stricken. This was duly carried out in October 1923.

Below: A-4062 in a spot of bother. It appears that the hull has sprung a leak and the machine has been beached. This does not look like the incident of 13 January 1919, when, with a crew of Ens Garrett B. Linderman Jr (p). Student officers CQM(A) Geroge Alfred Hero; CQM(A) Harry Cordery Moore & CQM(A) Robert Alston Brant; MM2c Orville Roy Bolton; MM1c Ivan Edward Quinlan, the boat crashed in Escambia Bay due to porposing and stalling to a total wreck. The cause was held to be glassy water. The boat was written off with 61:35 hours since 19 October.

Curtiss H-16 N4060 assigned to Felixstowe; it had a balanced rudder and other control surfaces.

Curtiss H-16 of No.257 Sqdn, RAF assigned to NAS Dundee in September 1918.

Curtiss H-16 N4072 of No.257 Sqdn, RAF assigned to NAS Dundee.

Above: An H-16 at Queenstown, Ireland, 1918. (AHT AL-1171-018)

construction and how the prototype overseas had performed entirely satisfactorily.[17]

US crews also flew British H-16 as well as F.2A boats. N4067 was delivered to the USN at Killingholme on 20 July 1918. Ens Benjamin Lee, USNRF, lost his life when this boat crashed in the River Humber during a delivery flight from Killingholme to Dundee on 28 October. Lt (jg) R.H. McCann, USNRF and Ens G.S. Hodges, USNRF, dove overboard under the burning wreckage from the rescue launch in their endeavours to rescue to seaplane's crew. The saving of the second pilot, Ens J. Garrison, was solely due to their efforts. Ens Hodges brought an unconscious Garrison through the burning wreckage to the rescue boat, keeping him under water intermittently and with his own body protecting Garrison from the flames. Hodges suffered severe burns due to his actions. Lt McCann dove repeatedly in and about the wreckage and did not cease his efforts until all hope of rescue had passed.

British H-16 boats N4062, N4067, N4068 and N4069 were used by the USN at NAS Killingholme.

The Aviation Matters Progress Report for October 1918 noted that the Curtiss plant in Buffalo had completed its contract for 74 H-16 boats and production of the F-5-L would commence in November. The Curtiss Engineering Corporation plant at Garden City would continue to supply H-16 at about one per week. F-5-L boats were being supplied by the Canadian Aeroplanes Ltd at the rate of five per month and from the Naval Aircraft Factory (NAF) at about three per week.

Many experiments were carried out on the USN H-16 boats during their service life. One, Bureau No. A-830, was built in an experimental pusher configuration with swept back wings.

Felixstowe F.2C

At the 12 June 1917, Meeting of the Air Department Progress Committee, Cdr Longmore reported that the trials of the F.2C were being carried out at Felixstowe. The machine was to go to the Isle of Grain for testing. The F.2C was described as being similar to the H-12 "but has a different hull which is not quite so efficient as a hydroplane and has been found to be considerably better for landing as it has not the same inclination to leave the water again on being landed as the H.12 hulls have."[18] The F.2C had a hull of lighter construction and greater volumetric size than the F.2A and F.3, its proportions seemingly to anticipate the F.5.[19] The hull had steps of revised design and the forebody's contours were different from those of the F.2A. The pilot's occupied an open cockpit. This was due to the fact that they had very little rear view in the canopied H-12 and F.2A rather than the usual answer that pilot's preferred an open cockpit. The front gunner's cockpit

H-16 Specifications

Source	1. N4060	2. British H-16	3. British H-16	6. British H-16	4. US H-16	5. US H-16
Span Upper	95 ft 1 in			95 ft 1 in	96 ft 6¾ in	96 ft 0$^{13}/_{16}$ in
Span Lower	69 ft 0 in				68 ft 11$^{3}/_{8}$ in	68 ft 11$^{3}/_{8}$ in
Chord	7 ft 0½ in				84 ft $^{49}/_{64}$ in	7 ft 0$^{49}/_{64}$ in
Gap					96 $^{9}/_{16}$ in	8 ft 0$^{9}/_{16}$ in
Incidence					4°	4°
Dihedral					1°	1°
Length	45 ft 7 in			45 ft 7 in	46 ft $^{15}/_{32}$ in	46 ft $^{15}/_{32}$ in
Height				18 ft 6 in	17 ft 8¾ in	17 ft 8¾ in
Aerofoil					RAF 6	
Tailplane Incidence					+1°	2°
Areas in Sq. Ft.						
Wings Upper without ailerons					616.2	601.0
Wings Lower					443.1	432.0
Ailerons					131	131.0
Wings total	1,200	1,200	1,108		1,190.3	1,164.0
Horizontal Stabilizer					108	108.0
Fin					31.1	34.0
Elevators					58.4	58.4
Rudder					27.9	27.9
Non-skid					24	
Capacities in Gallons						
Fuel	276					235
Oil	12					12
Water	2 x 4½					

Source:
1. Report NM200 on N4060.
2. Table "Flying Boats and Ships Aeroplanes," in *Flight*, 22.04.1920, P.441.
3. Table of Performance of Seaplanes & Flying Boats, 31.10.1917. Copy in RAFM J.M. Bruce Collection Box 26.
4. Curtiss Aeroplane & Motor Corp, NY, Catalogue - H-16-A specification sheet.
5. USN H-16A Handbook.
6. *Record of Performances of British Seaplanes 1917–Dec'31 1918*. Copy in TNA AIR1/708/27/11/02.

Notes:

* **Military Equipment:** 5 Lewis guns; 10 Ammunition trays; 1 Scarff mount; 2 outrigger mounts; 1 gooseneck mount; 1 pilot directing bomb sight, 1 bracket for Wimperis high altitude sight. 2 bombs with gear.

** **Engines, Etc.** Motors; Water; Radiators, Propellers; Gas pumps; Hand pumps; Gas tanks; Gravity tanks plus piping; Sil tanks plus pipping; Spark & throttle; Tachometers & shafting; Oil pressure gauge' Thermometers' Gas gauges' Tools & spares; Oil thermometers; Hand cranks; exhaust manifolds.

*** **Electrical Equipment:** Batteries; Lights; Switchboard; Wiring; Intercommunication set; Searchlight.

+ **Construction & Repair Equipment:** Bilge pumps; Sea anchor; Air-speed indicator; Inclinometer; Fire extinguisher.

++ **Navigation Equipment:** Compass; Watch; Altimeter; Chart board; Flags; Food & water; Pigeons; Binoculars.

+++ **Medical:** First aid kit.

H-16 Specifications (Continued 1)						
Source	1. N4060	2. British H-16	3. British H-16	6. British H-16	4. US H-16	5. US H-16
Weights in Pounds						
Crew 4	720	720	540	720		720
5 Lewis Guns, mounts, ammunition	255					
Bomb gear for 2 x 230-lb bombs	40					
W/T & Miscell.	177					
Total military load	1,192					
Empty	7,363	7,363	6,351	7,363	6,956	
Hull (inl soakage)						2,288
Wings, struts, tail, etc.						2,014
Military load less crew	472	472	467	472		850*
Engines, etc.						2,714**
Electrical Equipment						57***
Construction & Repair Equipment						75+
Navigation Equipment						39++
Medical						5+++
Radio						160
In flying trim empty (actually weighed)	7,835					
Fuel	2,001					
Oil	114					
Fuel & Oil		2,115	1,734	2,115		1,554
Gross	10,670	10,670	9,092	10,670		10,476
Speed						
Sea level			78			
at 2,000 ft	85½ kts	85.5 kts		85.5 kts		
at 5,000 ft	84½ kts					
at 6,500 ft		83.5 kts				
at 10,000 ft	80 kts	80 kts		80 kts		
Max						95 mph
Min						56 mph

H-16 Specifications (Continued 2)						
Source	**1. N4060**	**2. British H-16**	**3. British H-16**	**6. British H-16**	**4. US H-16**	**5. US H-16**
Climb Time to						
2,000 ft	3 min 40 min	3.7 min	4.0	3.7		
5,000 ft	10 min 20 sec					
6,500 ft		14.6 min	16.0	14.6		
10,000 ft	28 min 0 sec	28.0 min	36.0	28.0		
Ceiling in ft	12,500					
Endurance	About 6 hrs		5 hrs			
Engine	RR Eagle VIII		250 RR Mk.IV	RR Eagle VIII		Liberty
Date Trial Report		**05/18**	**13.01.17**	**05.1918**		

Above: H-16 Bu No. A-4061 with a Curtiss N-9 trainer overhead. A-4061 was reported as 100% complete in October 1918 and shipped the 16th to the Brooklyn Naval Aviation Storehouse but was diverted to Pensacola with A832-836 and A4060-4062. On 24 December 1918, it was on an instruction flight from Pensacola and at 1,500 feet when the port airscrew shattered causing a forced alighting near entrance to East Bay. It drifted ashore and was secured. On 9 May 1919, on a return from Mobile, the boat flew over Pensacola base at low altitude. The crew was Ens W.H. Robrack (p); Ens M.D. Nolan (assist p); Ens B.L. Chase (navigator); Ens F. Lamb (passenger); MM1c T.A. Hudson. Robrack was suspended from flying for one week as a result. On 11 October 1920, the port engine was damaged when the insulation wiring burned away causing the distributor head to burn. The upper engine panel, intermediate panel and lower intermediate wing panels burned. The incident was caused by a broken gas lead causing the motor to back fire. The flying boat was overhauled and the engine repaired. On 26 April 1921, Pensacola reported that the condition of the hull was generally bad, with timbers water soaked and rotten in the bottom; planking was coming loose from the hull; interior fittings were very rusty. The damage was incident to service and it was recommended that a spare hull be substituted and minor repairs be carried out to the wing surfaces. Engines 0508/966 and 02247/3576 were to be overhauled. On 21 May it was reported with a new hull, and the minor repairs to wing surfaces were nearly complete. The following month the new hull suffered in a forced alighting. The upper left hand longeron was broken, and other damage caused by the failure of Engine No.308/572. The boat undertook a test flight at Pensacola on 24 September before being shipped by air to Hampton Roads with A-850 and A-3500. Hampton Roads reported on 12 January 1922, that A-4061 was in commission but was in excess of allowance, taking up manpower and space. A request to place the machine in storage was granted. On 1 October 1923, it was sold without engines or instruments for $45.

H-16 Bureau Number Allocation				
Bureau No.	**Number**	**Contract**	**Manufacturer**	**Notes**
A-784 to A-799	16	SOC-626 Wash CR-121; SE-250; Bu-18	Curtiss A & M	
A-818 to A-867	50	C-33181 CR-26; SE-258; Bu-18	Curtiss Eng	
A-1031 to A-1048	18	C-33193 CR-184; SE-298; Bu-18	Curtiss A& M	
A-1049 to A-1098	50	RNAF-Z-1	NAF	
A3459 to A3558	100	RNAF-Z-1	NAF	A3508 & A3582 used as spares, Norfolk.
A4039 to A4078	40	C-33193 CR-184; SE-336; Bu-18	Curtiss A & M	

H-16 Serial Allocation			
British Serials	**Number**	**Contract**	**Notes**
N1890–N1949	60	AS6731 (BR50)	Serials cancelled. Renumbered N4890–N4949.
N4060–N4074	15	N1387-1137W	All delivered without engines.
N4890–N4999	110	AS6731 (BR50)	From N4900 directly into store. All scrapped except N4902 and N4905 that were gifted to Canada. N4950 to N4999 cancelled.
Note: H-16 boats in the N4060 series were being delivered at the same time as those in the N4890 series.[21]			

Above: H-16 A-836 on the ramp at Pensacola. It was usual to start the engines before lowering the boat into the water, and it is therefore likely that this photograph shows the recovery of the machine. Navy Grey overall. This boat was lost on 10 January 1921 at Pensacola. The crew were: CMM(A) J.W. Utley; CMM(A) J.E. Rawlings (student). Held to have been poor judgement in recovering from one engine turn resulting in a bad skid; landed with one wing low to a total wreck except for Engines 0988/2154 and 0751/1595 that were not damaged. Fortunately there was no injury.

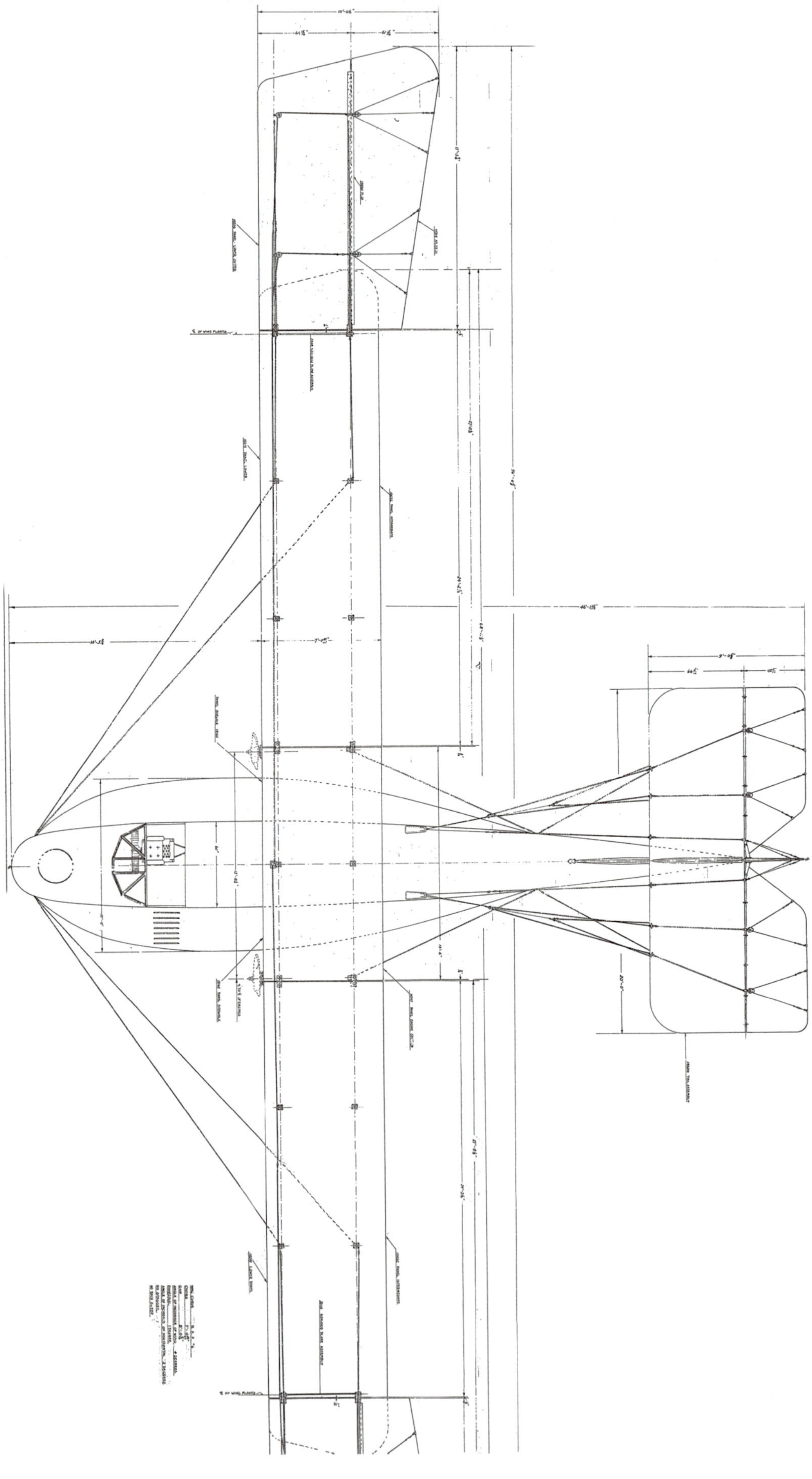

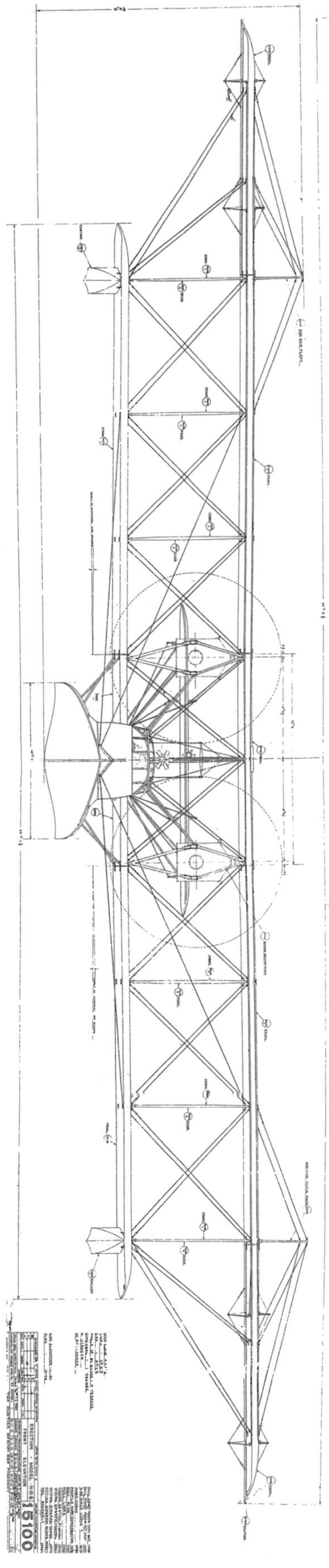

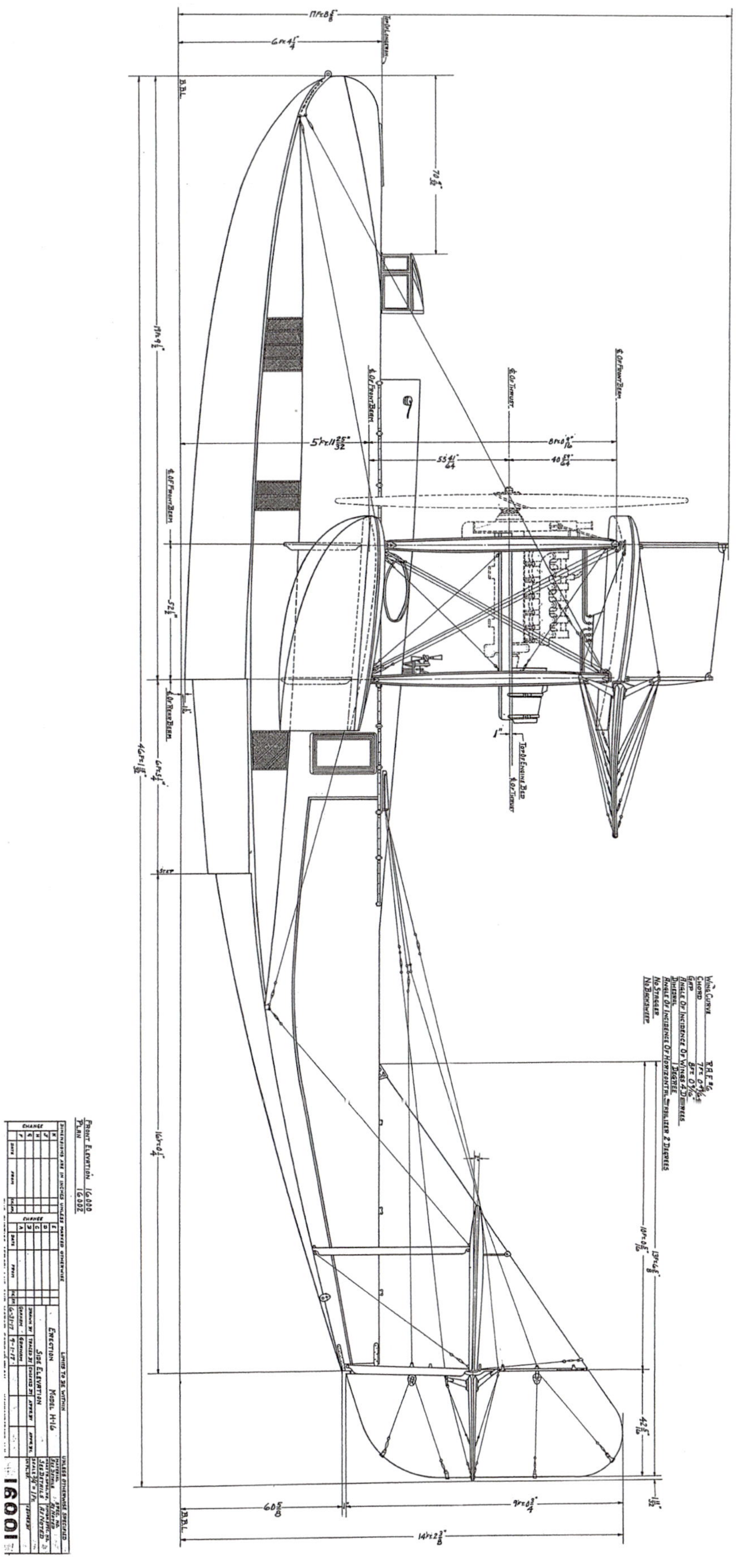

Facing Page, Top: The H-16 did carry out patrols from stations in the UK. A rear view of K-11 from Killingholme shows the twin 520-lb bomb load of the type. 24 October 1918.

Above: A front view of a Killingholme bombed up H-16, 24 October 1918.

Right & Facing Page, Bottom: Detail of the Skeleton Tubular 520 Lb Mk.II(F) bomb carrier and an H.E.R.L. 530 lb Mk.1/N/a light case bomb on a Killingholme H-16, 24 October 1918. (The N indicated Naval use).

Above: H-16 A-3479 shows the application of the last two numerals of the Bureau No. to the bow as the aircraft code number. This boat has the balanced rudder. A-3479 had a hull built by L.C. Seabury Co. This machine has a confusing history. A-3479 was reported as 100% complete in August 1918 and shipped to Pensacola on the same month by rail. Pensacola reported it on station on 4 November. It did not last long as on 5 December it was reported sinking off Town Point. A barge was dispatched but failed to tow the boat ashore before it sank. The cause was held to be due to landing on glassy water which crushed the hull bottom. There were no injuries in the incident. However, other sources state that it was railed to Pensacola on 9 December 1918, and was in service during 1921 and 1922. Given the colour scheme and markings in the photograph this seems most likely the correct history. Whether the original boat was recovered or the entry is incorrect is unknown at this time. A-3479 was eventually stricken on 16 June 1922, owing to general poor condition, the wing fabric being poor, the hull old and leaking. The engines, 02218WR/3388 and 05506WR/4761, were to be overhauled.

Left: Assembling or disassembling an H-16 in France. The aircraft code ST-16 indicates it was assigned to Saint Trojan NAS. A far as can be ascertained from surviving documents no H-16 flew offensive patrols from French stations. (AHT AL0340-103)

Above: Curtiss H-16 boats A-836 and A-3479 on the hardstand. The Bureau Nos. marked on the hull sides indicates that these are US based H-16 boats, this photograph being taken at Pensacola Naval Air Station. Note how the last two numerals of the Bureau Nos. are painted on the bow. There appears to be two Curtiss HS-2L boats in the background but it is only one machine caught by the time exposure so that the Bureau No. A-2082 appears to be on the front and rear hull. This machine bears the "hat in the ring" insignia of one of the squadrons that operated from Pensacola. N-9 training biplanes are further to the rear.

A-836 was delivered to Pensacola in September 1918 in a damaged condition thought due to the crates hitting overhead bridges on the Atlantic Coast rail line. It was stricken but obviously repaired and back in service by at least 9 December 1918. This boat was crashed to a total wreck on 10 January 1921.

A-3479 was on NAS Pensacola on 4 November 1918, and survived to be written off due age and condition on 12 May 1922. A-2082 was on Station in October 1918, and was written off on 14 April 1924, due to condition due terms of service.

Right: A-867 was 100% complete in January 1919. On the 27th it was shipped to the NAF for storage, arriving in company with A-845. It was sstill in storage in July 1923 when it was noted as requiring reconditioning. In May 1925 it was allocated to the USS WRIGHT. On 14 August 1926, it was received back at the NAF. In March 1927 it was recorded as with VJ 1. On 3 May following year there were signs of rot on lower ends of the centreline, struts, etc., and the boat was to be stricken. This taking place on 27 June, with a TFT of 476.75 hrs. A-867 can be seen in the photograph to have all the late modifications to the type. The cockpit is the open type, there is a balanced rudder and the rear of the hull has ply sides with strengthening ribs. The Bureau No. is now carried in small numbers on the fin. The hull appears to be a colour other than Navy Grey.

Above & Below: Views taken on an H-16 in flight. The dual control wheels are visible through the cockpit canopy.

Facing Page: Ens W.A. Gus Read demonstrating an experimental machine gun mount on the rear of an H-16. The awkward gallows mount for rear protection is operated by the rating in the side window. The wing walks on the root of the lower wing and the cloth wrapped engine support struts are noteworthy.

Above: NAF constructed F-5-L A-3560 shows the same rear gun mount as on the H-16. It is not known if the ring could slide along the rails at the side of the large rectangular opening in the hull top decking. Note that the hard stand at the top of the slipway is timber planked. Reported "Ready. Loaded on car" on 28 August 1918, A-3560 was shipped to Hampton Roads by air on 9 September arriving the following day. On 12th it attempted to get off with a gross load of 13,094 lbs for a test. On 18 September the hull was reported as crushed on right side covering an area about 2 sq ft. Wires were loosened on the tail. The aileron control fitting and pulleys were torn out of the center top plane. Entire upper left wing from the engine section out, including the overhang complete with aileron and fittings was damaged beyond repair (DBR). The machine was to be repaired on Hampton Roads station. Get away and propeller tests expected to continue on 8 October. On the 19th it had another accident. The right aileron and horn were DBR; the left aileron and horn were badly damaged; while the upper load wire fitting on the intermediate wing were broken when the plane collided with a light beacon while manoeuvring for pictures. Repaired it was with the Hampton Roads Experimental Department in January 1919 when it flew to Washington and return to test a radio transmitter. On 3 May 1919, as a result of alighting in a heavy sea the machine was lost but the engines were salvaged. A-3560 was stricken on 9 June with a TFT of 5:58 hours.

Facing Page, Bottom: A-839 was built with its engines installed in pusher configuration and the wings were given a large sweepback. Also noteworthy are the stencils along the side of the walkways across the fins of this boat. Early US based H-16 boats appear to have had the gooseneck spigot mount for their front machine gun. Reported as 100% complete in October 1918 at the Curtiss Garden City plant. On 2 November underwent a preliminary trial when flown by Lt McCullough at Curtiss Garden City. It rose from water twice but only for short hops. Found to be slightly nose heavy the boat was taxing for a third flight when the propeller split damaging the port side walk panel and other panels. In December it was recorded as now ready for trials at NAS Rockaway. While there it was stored outside due to lack of space. Flown to Hampton Roads on 13 January 1919, it had special auxiliary rudders placed within streamline of the propellers as shown in the photo. In February it was laid up for repairs. On 26 October 1920, Hampton Roads recorded that the hull was rotten and beyond repair due to the length of service. The wings could not be used on the standard H-16, and it was recommended it be stricken. This was carried out in November. There appears to have been no advantage to the pusher configuration and the machine was little used.

Right: The front gun position on an H-16 with US produced Scarff ring mount for the Lewis gun.

Above: A-3500 being shown off to a couple of young ladies.

was farther back from the bow and there were no waist gun positions. Powered initially by two 275-hp Rolls-Royce Eagle II engines, these were later replaced by 322-hp Eagle VI engines. Official performance trials were held on 23 June 1917, and although performance was slightly better than the F.2A, there was no chance of it being placed into production and upsetting the F.2A program that was then underway, only the prototypes N64 and N65 were built.

N64 and N65 were sent to Felixstowe where they joined the War Flight. Porte led a patrol of five flying boats from Felixstowe in N65, his latest experimental boat, on 24 July 1917. His pilot was *Queenie* Cooper, the test pilot. This was the first time that Felixstowe had been able to field this many boats at once. They spent a long time getting into their correct position for take-off. In future this was to become common and they were got away easily.

Discovering a submarine on the surface, the patrol attacked.[20] Three of the boats dropped bombs, including N65. The other two boats stood by but the submarine appeared to be finished. This submarine was incorrectly reported as UC-1.

While on a Beef Convoy patrol in late September 1917 *Pix* Hallam and Watson in N65 sighted a submarine and they immediately attacked. The F.2C was equipped "with a gadget for dropping bombs by compressed air, which, according to its proud inventor, was to supersede the good old way of dropping them by pulling a Bowen wire."[21] Unfortunately a good attack was ruined when the bombs hung up. The device hissed as compressed air escaped but the bombs refused to leave the racks. A nearby destroyer then attacked with depth charges.

N65 was written off after it was holed and sank at the Isle of Grain on 13 March 1918.

N64 was converted into an F.3 about October 1917. It was used for patrols and for experiments.

Felixstowe F.3

The F.3 was designed to meet the need to carry a larger bomb load as the 230-lb bomb was considered no longer satisfactory for attacking submarines. A Progress Committee Report of 12 June 1917, noted that all new "Americas" now on order would be of this type.[22]

The F.3 actually preceded the F.2C. The F.3 had the same large overhang of the upper wing and king-post structure above the interplane struts. Its hull was three feet longer than the F.2A and the wing were of greater span and chord. These differences are not very noticeable in photographs and if the serial is not visible the only way to tell the difference is that the F.3 had a

Above: The solitary F.2C, N65, on the hardstand area in front of the Felixstowe hangars. Together with 8676 and 8689, N65 was credited with helping sinking U-Boat *UC-1* on 24 July 1917, with Wing Cdr John Porte at the controls.

Below: The end of N65.

F.2C Specifications				
Source	1	2	3	4
Length			46 ft	46 ft 0 in
Span			95 ft	95 ft
Height				17 ft 3 in
Wing Area	1,136 sq. ft.		1,136 sq. ft.	
Weights in Pounds				
Empty	6,768		6,768	
Fuel & Oil	2,350	1,766*	2,350	
2 Lewis Guns & Scarff Ring Mountings		200		
Bomb sight & Fittings		40		
Military load	402		402	
Crew	720	540	702	
W/T & Sundries		171		
Dead weight		886**		
Gross	10,240		10,240	
Speed				
at 2,000 ft	85 kts		98 mph	
at 2,100 ft		85 kts		
at 3,500 ft		84 kts		
at 6,500 ft	82 kts		94 mph	
at 7,180 ft		80 kts		
at 10,000 ft	78.5 kts		91 mph	
Climb				
to 2,000 ft	4.5 min		4min 50 sec	
to 3,000 ft		7 min		
to 6,500 ft	18.3 min		18 min 20 sec	
to 10,000 ft	38.0 min		38 min 00 sec	
Ceiling	13,000 ft		10,300 ft	
Endurance	6½ hrs	5 hrs***		
Engine	275 RR Mk.II			
Date Test Report	23.06.17		23.06.17	

* 220 Gall fuel & 16 Gall oil.
** Dead weight in lieu of 125 Gall petrol.
*** Extra deadweight carried on trial brings endurance up to 7½ hrs.

Source:
1. Table of Performance of Seaplanes & Flying Boats, 31.10.1917. Copy in RAFM J.M. Bruce Collection Box 26.
2. F.2.C. Large America Boat. 22nd Weekly Report of the Controller, Technical Department for W/E 18 July 1917. Copy in RAFM J.M. Bruce Collection Box 25.
3. J.M. Bruce data.
4. *Record of Performances of British Seaplanes 1917–Dec'31 1918*. Copy in TNA AIR1/708/27/11/02.

Above: The end of N65. From the damage it appears that the bottom of the hull was torn out, a too often common occurrence with the *Large America* boats.

slight recess in the leading edge of the upper wing immediately behind the airscrews.

The F.3 had a typical Porte hull and was planked the same way as the F.2A. The fins were planked with three-ply birch. On later examples double diagonal planking was used. The floors of the F.3 were rebated into the keelson on either side and caused the planking to spring and the boats to leak badly. Intermediate timbers were introduced to try and overcome the problem. Sir A Robinson recalled that on

> *at least one station, where they often had to be bounced off a long swell, it was found that the planking was apt to open up from the keelson. A new boat on arrival would very probably be dismantled and the petrol tanks removed, to permit a series of oak knees to be put in to strengthen and hold together the bottom planking.*[23]

The F.3 could carry four 230-lb bombs while the F.2A could only carry two. This may be the reason that more F.3 boats were ordered than F.2A boats. Lt (later Major) J.D. Rennie was Porte's Chief Technical Officer at Felixstowe and who assisted him in the development of the successful Felixstowe flying boats stated post-war that the F.3 should never have been put into production.

The prototype, the converted N64, was powered by two 320-hp Sunbeam Cossack engines. Production machines had Rolls-Royce Eagle VIII engines.

The F.2A boats went to Felixstowe, Yarmouth, Dundee, Calshot, and Killingholme, until that latter station was taken over by the USN and they began to receive Curtiss H-16 boats. The F.3 on the other hand went to Cattewater, and the Scillies, and to Houton Bay in the Orkneys. The reason for this was simple. The F.3 used the same Rolls Royce Eagle VIII engines that the F.2A used, was heavier and slower and less manoeuvrable and was unable to engage Zeppelins or German seaplanes. The F.2A went to those stations where reconnaissance and fighting was the most important, while the F.3 with their greater fuel capacity and bomb load went to stations where aerial opposition was less and hunting U-boats was more important.

The first operational flight of a F.3 was in July 1917. The type came in for modifications at the stations as had happened to the F.2A. The tail fabric would be replaced with ply sheeting. Sometimes the double bottom planking was put on the wrong way around so that the water flow was across the outer skin of planking and not along it. The bottom would be removed and replanked, a not infrequent activity according to Sir Austin Robinson.

A report on a Phoenix built boat, N4400, noted that:

1. The mahogany outer planks of the hull bottom aft of the step were buckling outwards, due to swelling. The planks were about 4½ inches wide and were only secured along the joints, no intermediate copper clenched nails being employed.

T.5. LARGE AMERICA F.3.
(Two 320 H.P. Rolls-Royce.)

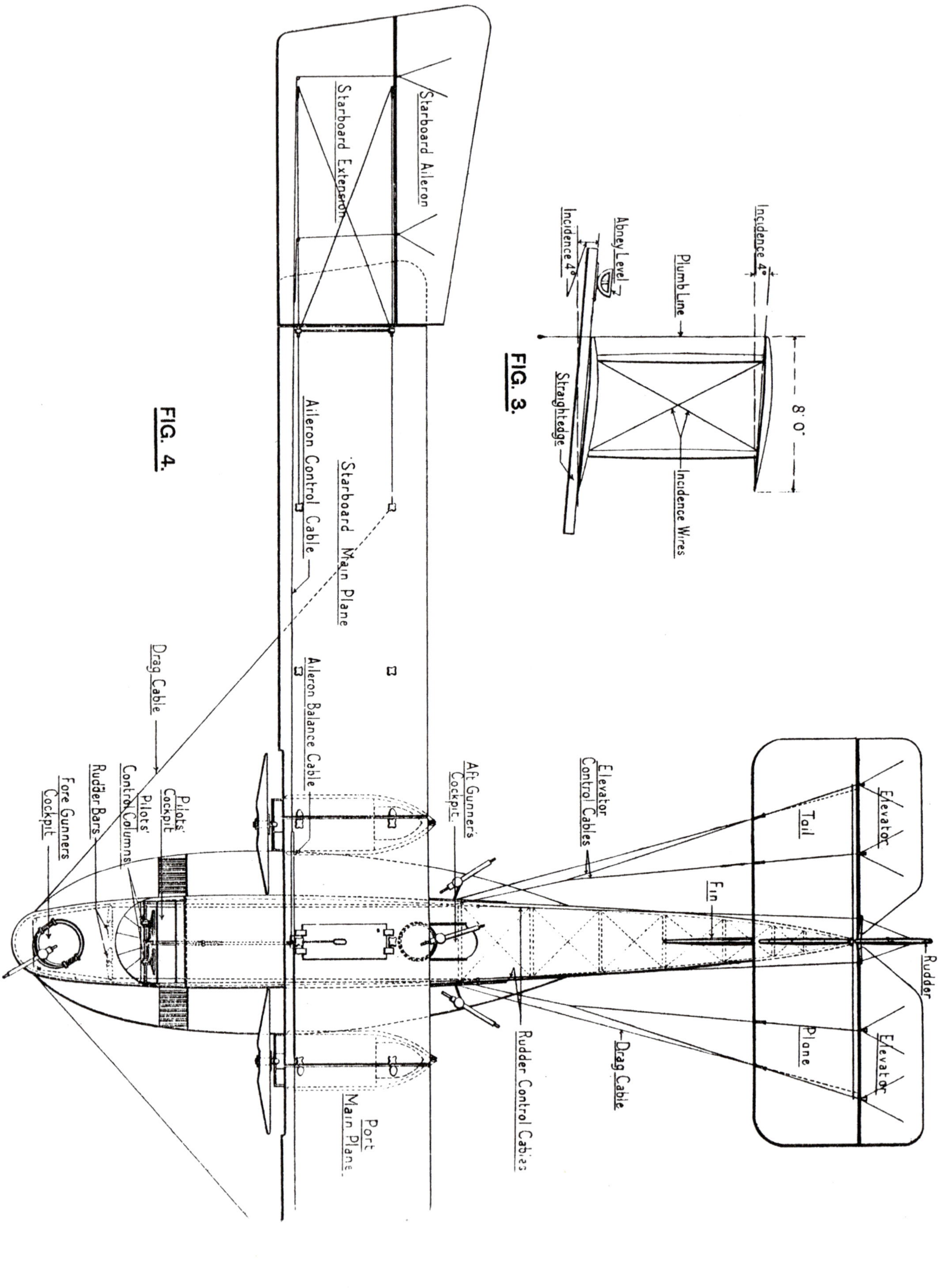

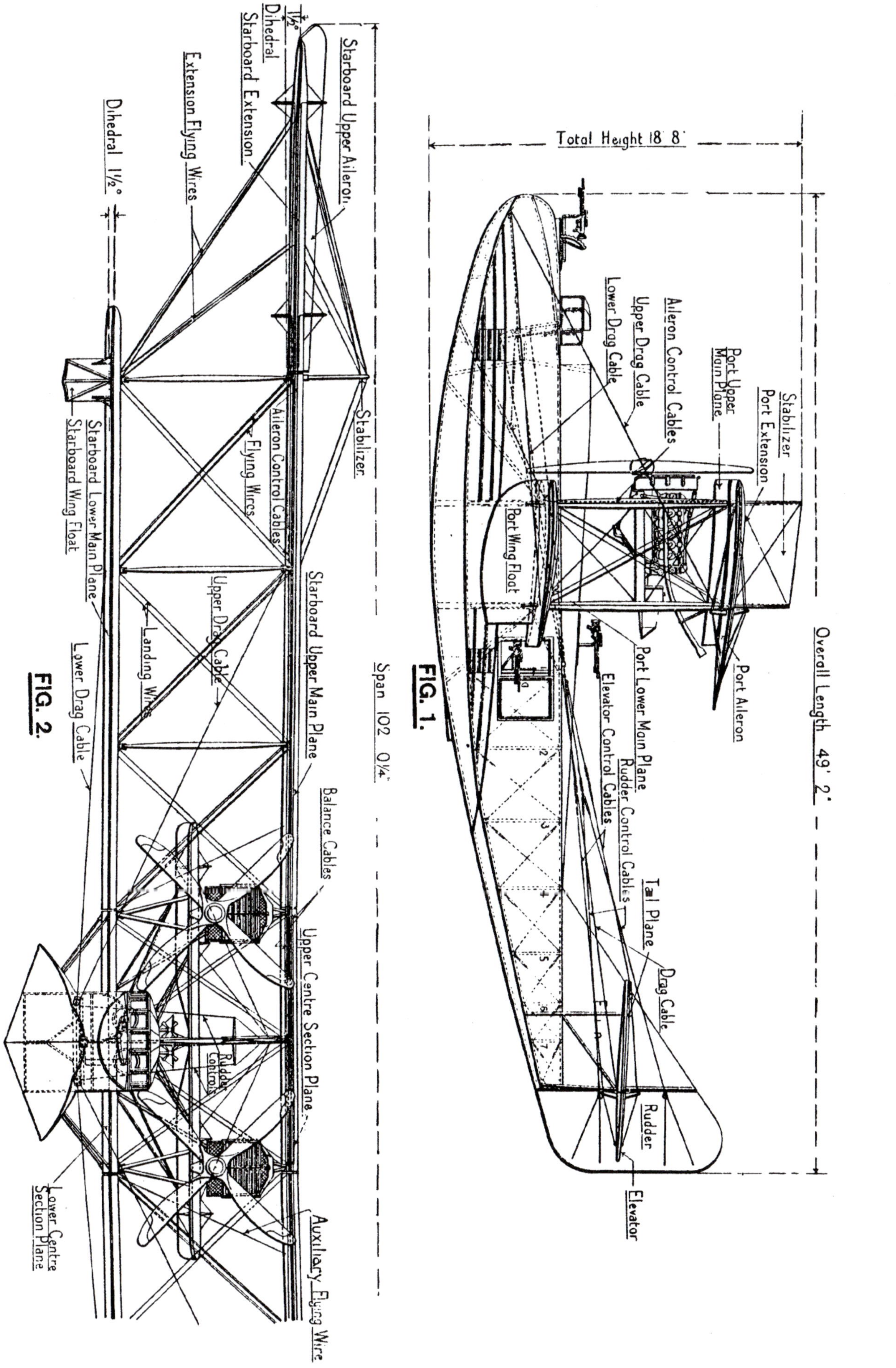

LARGE AMERICA F.3.
(Two 320 H.P. Rolls-Royce.)

Above: This F.3 has its dazzle painted hull in broad bands of colour.

Above: "Just Away." Launching F.3 serial N4193. The fabric surfaces appear to be V.84 aluminium doped. N4193 was with No.230 Squadron in 1922–1923. Stationed at Calshot in 1924–1925 it was reconditioned by Shorts.

Above: TN4404 appears to have the upper section of the hull above the fins finished in the same colour as the fabric surfaces. Note the stencil under the tailplane "Phoenix Bradford No.204". The arrow markings on the hull indicate where to position the docking trolley. Phoenix personnel pose with the boat at Brough before delivery. N4404 was of a batch of 30 boats built by Phoenix in early 1918. Delivered to Houton Bay in April it was written off when it struck a submerged object on takeoff on 23 May and was deleted the following June.

Below: Two F.3 boats beached near their unlucky companion that has capsized. The anti-fouling bottom finish is to be noted.

Above: F.3 N4415 on its beaching trolley parked on the hardstand at Tresco station. This photograph is dated as April 1919. (Another photograph of this machine appears on P.56 of *U-Boat Hunters – Cornwall's Air War 1916–19*, by P. London.) This machine was recorded at No. 234 Sqn at Tresco on 11 October 1918, prior to that it had served with No.238 Squadron at Cattewater being recorded for the W/E 22 August. The main hangar was located just behind the camera. Tresco was to have 12 of the F.3 boats but this number was never achieved.

Below: An F.3 on its handling trolley. The lack of a serial number is unusual.

Above: Beaching a Felixstowe F.3 at dusk.

Above: Transporting an F.3 hull. On this hull the washboard and nose are a dark colour (black) while the fins and hull bottom appear to be varnished. Note the abbreviated type cockpit canopy.

Above: An F.3 in front of permanent hangars.

Below: F.3 N44XX appears in this photograph to have the hull painted the dark upper-surface colour over all surfaces.

Above: Details to be noted from this photograph are the type of gun ring, bomb sight, radiator and bomb racks, the position of the wind driven fuel pumps behind the cockpit. The nose and hull bottom appear to be black. The bottom wing engine section next to the hull where the rating is standing, would have had a ply walkway.

F.3 Specifications

Source	1	2	3 Normal	3 Overload	4	5	6
Span Upper		102 ft 0¼ in				102 ft	102 ft 0 in
Span Lower		74 ft 2 in					
Chord		8 ft					
Incidence		4°					
Dihedral		1½°					
Length		49 ft 2 in				49 ft	49 ft 6 in
Height		18 ft 8 in				18.75 ft	
Incidence Tailplane		2°					
Wing Area, sq.ft.			1,430	1,430	1,429	1,413	
Weights in Pounds							
Empty	8,270		7,958	7,958	8,035	7,958	
Fuel & Oil	2,455		2,096	3,142	2,790	9,095	
Military load	Nil		1,461*	1,461*	-	1,461	
Crew	720		720	720	720	720	
Gross	11,625		12,235	13,281	11,725	12,235	
Speed							
At 400 ft	88.5 mph						
at 2,000 ft			79 kts	78 kts	78.5 kts		
at 6,500 ft			74.3 kts	75.5 kts	76 kts		
at 10,000 ft			—		72 kts		
Climb							
to 2,000 ft	3 min		5.4 min	7.8	4 min		
to 6,500 ft	27 min		24.0 min	41.0	20.0	24 min 20	
to 10,000 ft	60 min		—		58.0		
Ceiling in ft			8,000	6,000	11,000		
Range						880 miles	
Endurance			6 hrs	—		11 hrs	
Engine					320 Sunbeam	Eagle VIII	
Engine	320-hp Sunbeam Cossack	345-hp Rolls Royce					
Date Trial Report	09.02 & 19.04.1917		04/18	04/18	19.04.1917		

The F.3 generally carried the same armament as the F.2A but could carry four 230-lb bombs against the F.2A's two.

Sources:

1. J.M. Bruce data.
2. F.3 Rigging Notes & Aeronautics, 29.01.1919.
3. Table "Seaplanes & Ships Aeroplanes" in *Flight*, 22.04.1920. P.441.
4. Table of "Performance of Seaplanes & Flying Boats", 31.10.1917. Copy in RAFM J.M. Bruce Collection Box 26.
5. Post-war advertisement for the "Short F.3 Flying Boat" made by Short Bros., Ltd.
6. *Record of Performances of British Seaplanes 1917–Dec'31 1918.* Copy in TNA AIR1/708/27/11/02/

F.3 Serial Allocation					
Serials	**Number**	**Contract**	**Manufacturer**	**First Delivery**	**Notes**
N64	1	BR117	Felixstowe NAS		Completed as F.2C, converted to F.3.
N1950–N1959	10		AMC Ltd		Cancelled.
N2160–N2179	20	AS11426 (BR22)	Handley Page Ltd.		Renumbered N2160–N2179.
N2305–N2307	3	BR80	May, Harding & May.		Cancelled.
N2310–N2321	12	BR86	Malta Dockyard		Renumbered N4310–N4321.
N2400–N2449	50		Bristol & Colonial Aeroplane Ltd		Ordered 16.07.17, with 320-hp Rolls Royce engines. Cancelled, serials reallocated.
N4000–N4037	38		Short Bros Ltd.	11.07.18	All delivered. N4019 became G-EAQT.
N4100–N4149	11	38a/1090/C & AS4499/18	Dick, Kerr & Co Ltd.		Contract for 11 F.3 & 39 F.5.
N4160–N4179	20	AS11426; AS30303 & AS30620 (BR22)	The Phoenix Dynamo Manufacturing Co Ltd		All delivered. N4177 became G-EBDQ.
N4180–N4229	11	AS44496/18 (BR348)	The Phoenix Dynamo Manufacturing Co Ltd		Contract for 11 F.3 and 39 F.5. F.3 to N4190. N4191 F.3 or F.5?
N4230–N4279	50	AS13823 (BR72)	Dick, Kerr & Co Ltd.		Dick, Kerr's first F.3s. N4268–N4273 & N4276–N4277 no record of delivery.
N4310–N4321	12	AS14835 (BR86)	Malta Dockyard.	20.03.18	All delivered from March 1918.
N4360–N4397	38	AS14835 (BR. Adm1269)	Malta Dockyard.	05.09.18	N4388–N4397 cancelled.
N4400–N4429	30	AS30620 (BR199) & AS4496/18	The Phoenix Dynamo Manufacturing Co Ltd.		07.02.18. Grain test.

Note that aircraft were not built in order of serial. Some batches with late serials were completed by a manufacturer before a batch with lower serials that were contracted to the same company.

Above: N4410 has its engines and cockpit covered with canvas. Note that the wing in the background has a balanced aileron. This machine was at Houton Bay in July 1918, and at Dundee in January 1919.

Right: This Colourful F.3 has major damage to its fin and hull bottom. The fabric above the washboard at the rear of the hull has been removed.

Above: An F.3 at the factory ready for hand-over.

Below: N4232 in a bit of trouble on the slipway. In this photograph the difference between the dark rear hull fabric coloured surfaces and the washboard and fin can be clearly seen. This Dick, Kerr & Co Ltd built F.3 was one of a batch of 50 0rdered in May 1917, and was with 301 Flight Catfirth by August 1918. It served at Dundee and Houton Bay. It was deleted in the W/E 19 December 1918.

Above: Close up of an early F.3 shows details of the gun ring, cockpit canopy, ply walkways on the engine section of the lower wings, and the walkway over the fins.

Below: N4230 displays yet another variation to the colour schemes applied to the F.3. N4230 was the first F.3 of a batch of 50 contracted from Dick, Kerr and Co. The washboards are a much darker colour (black) than the pigmented doped fabric surfaces. This colour carried onto the fins, but does not extend for the full length of the fins as they are varnished at the front as is the ply sheeting of the forward hull. The nose of this boat is again this dark colour (black) and it carries across the front of the fins. N4230 undertook hydrophone trials at Shoeburyness on7 and 15 May 1918, and is illustrated in the experimental section photographs. In addition this boat took part in an attack on a submarine on 20 July 1918, when it dropped two 230-lb bombs on a U-Boat. It was still recorded at Houton Bay in January 1919.

Above: N4364 displays a long radio aerial. This machine appears a dark colour overall with the exception of the hull decking immediately behind the cockpit, however this may be due to the lighting conditions. The style of serial display is noteworthy. N4364 was one of those built at the Malta Dockyards by local artisans and was delivered to Calafrana in November 1918.

Above, Below, & Facing Page, Bottom: N64 in the guise of an F.3. Note the cockpit canopy and anti-skid fins on the engine support struts. As an F.3 at Felixstowe from October 1917, it was deleted in May 1918 after being reported as worn out the previous month.

Above: Front view of an F.3.

Below: Beached N4260 with damaged starboard wings and rudder. The engines had been removed by the time this photograph was taken. This Dick, Kerr and Co seaplane was delivered in August 1918, wrecked in transit to Cattewater and deleted in the W/E 31 October, apparently never having made an active patrol.

Above: There would be no excuse for using the wrong trolley for this F.3, N4258. One of 50 from Dick, Kerr and Co, this "dazzle" painted boat has the open cockpit and was with Felixstowe from September 1918 to at least January 1919. The crew in this photograph are unfortunately unknown.

Below: N4400 running up its engines at the Phoenix works in, from the apparel the civilian and military onlookers are wearing, what are windy and wet conditions. N4400 was first of a batch of 30 F.3 seaplanes (N4400-N4429). This machine was fitted with the Cooper "autoflare" landing stick and is probably that illustrated in the experimental section photographs. A number of other experiments were conducted with this boat and it was reconditioned by Short Brothers in 1920 and sold to the Portuguese Government in April that year, becoming C-PAON.

Above: Under these lighting conditions N4404 shows a different tone to the washboards. These two photographs highlight the difficulties of trying to determine colour schemes from black and white photographs.

Below: N4404 in the Phoenix colour scheme of all fabric areas given a pigmented dope finish (dark green or P.C.10?). In this photograph the washboards appear to be the same colour as the adjacent fabric. The stencilling under the tailplane reads: "PHOENIX BRADFORD, No.204." The white arrows on the hull would be locating points for the docking trolley. Note the fabric covered struts have wound cord bindings at their centres.

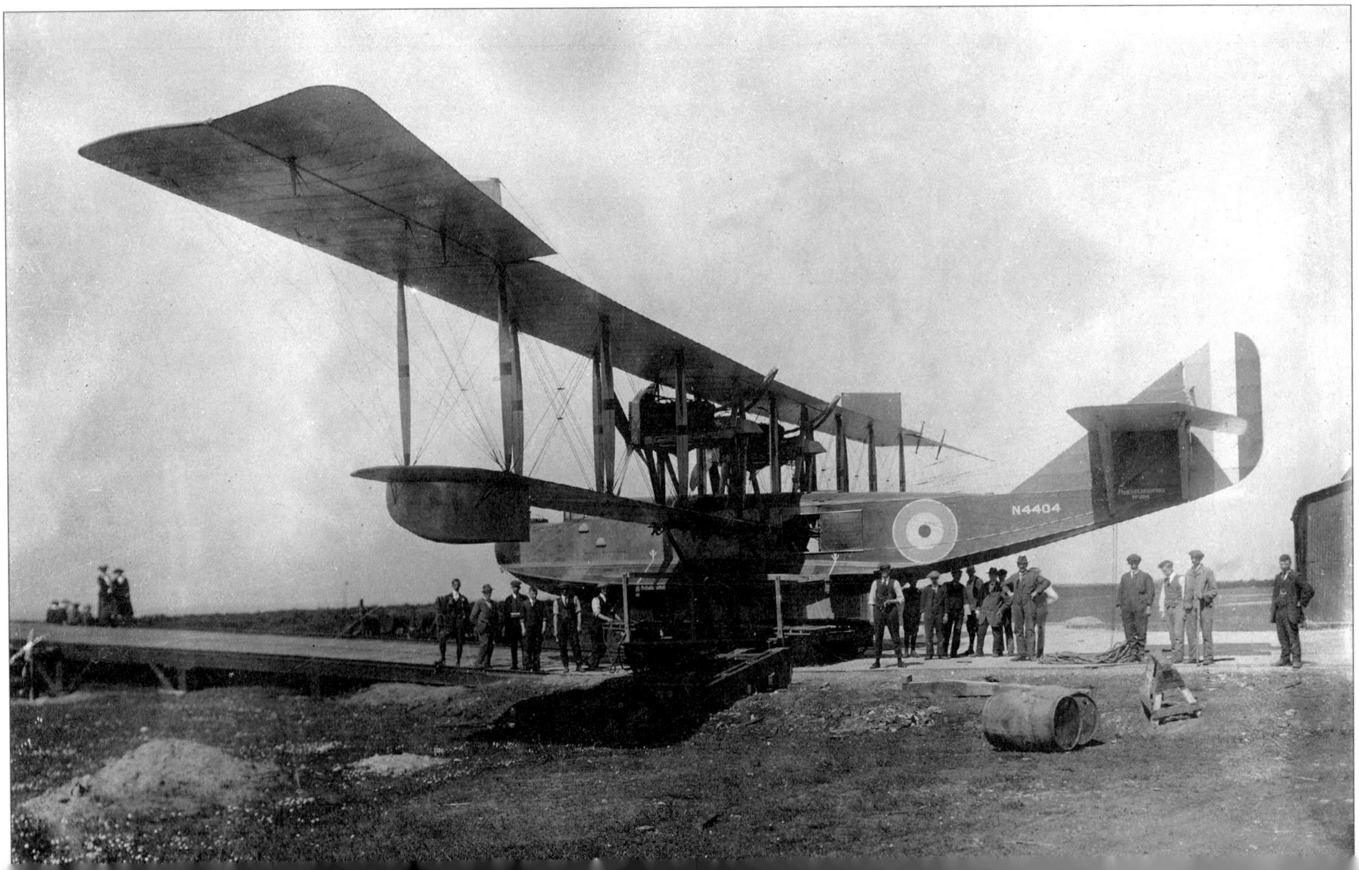

Above: Detail of the complex supporting sturts for the tail structure of N4406 in the Phoenix works, a boat from the same batch as those opposite. In this photograph it can be seen that the finish applied to the washboards was carried to the underside of the hull. 4406 was delivered to Houton Bay in June 1918, and was still there in the following October.

Below: N4404 on the water. Note the window on the port side of the upper hull under the wings. On this photograph the washboards appear to be a different colour from the adjacent fabric covered surfaces.

Above: N4409 is only three down the line from N4406 but is in a different colour scheme. All the hull fabric areas appear to be clear doped with black washboards and hull bottom. This seaplane was also delivered in June 1918.

Below: This photograph of N4410 bears close study. This F.3 carried the same colour scheme as N4409. The lower cockade is under the bottom wing. The wingtip float bears the stencil: "P3/T6/TB/030". The small dot on the outside of the float is the Phoenix Co's decal. A larger circular decal is on the side of the hull near the front of the boat, with a still large Phoenix on the bow. The stencil under the footstep reads: MAXIMUM LOAD NOT TO EXCEED 4250 LBS. The cockpit canopy with the window separating it from the main hull is well shown. Note how the top of the fins are clear varnished with the black giving an outline to the fins. N4410 served with No. 306 Flt from Houton Bay and was still there in January 1919.

Above: N4415 bears the same type of scheme as the earlier N4404. The serial numbers do not give the true order of delivery.

Below: N4413 under power. The rear of the hull is well down in the water. On the curved H-12 hulls suction was created and the Porte designed hulls overcame this fault.

Above & Left: The office of an F.3 flying boat.

2. Brass screws were freely used to fix the planking to the heavier framing timbers, and many were sunk too far into the outer planking. They should have been only slightly countersunk.
3. The footsteps were not correctly finished off to prevent water getting into the hull. "These should be boxed up on the inside of the hull to make the hull watertight."
4. The interplane flying and landing cables were not correctly finished.
5. An elastic adjuster "will be fitted to the cloche for balancing the elevators, to replace the present crude arrangement."
6. There was a great deal of sway between the hull and the main spars.

The report noted that the defects would be corrected at the station and suggested remedies for some problems.[24]

In October 1918 the Technical Department of the RAF wrote to the Admiralty noting that problems with the hulls of flying boats was due to "lack of sufficient transverse strengthening members and incorrect method of fixing the plank on the bottom

Above: Tresco Seaplane Station in November 1918. The station was established on the second largest island in the Isles of Scilly. The station was established in February 1917 and provided patrols for the Western Approaches. Two boats are on the hardstand area with a solitary Short floatplane.

Below: A dazzle painted F.3 rescuing the pilot of a Sopwith "Kitten", presumably a Sopwith Baby floatplane. (Ens Raymond L Atwood collection/Emil Buehler Library)

hulls." While modifications had been introduced some three months back great difficulty had been experienced with hulls in store and they had been issued to the various "super-structure Contractors" who were not hull builders and had neither the expertise nor facilities to undertake the modifications required. Lack of accommodation and lack of the necessary facilities to carry out the modification work on stored hulls was blamed for the slow rate of production of modified hulls from store and the issuing of unmodified hulls.[25]

The following report of 23 June 1918, from Houton Seaplane Station captures the problems present in the manufacture and operating of these large flying boats. Houton reported that none of its flying boats were fit for action as follows:

- N4245 Capsized on slipway in strong side wind.
- N4232 Sank on arrival and is being recaulked and fitted for service.

- N4406 Developed leaks in hull after 48 hours on the water. Sent to Houton from Stenness for repairs on 21 June.
- N4235 Sank on arrival and is being recaulked and fitted for service.
- N4403. Driven ashore in gale 13 June, bring repaired.
- N4230. Undergoing its first fitting out on arrival.

These were all F.3 boats. The Station also had Porte Baby 9810 that had been driven ashore in the 13 June gale. It had yet to be floated, a special slipway and cradle were being constructed for that purpose.

This unsatisfactory state was due to:

1. The station not being fully completed for service.
2. The machines being leaky and not ready for service when they arrive.
3. The bad weather in June.
4. A severe epidemic of influenza at Houton.

The majority of men on the station were recruits of one or two months service and had no experience of repairing or handling *Large America* seaplanes. Only one shed had been available and that provided accommodation for only one seaplane, and all fitting out and repair of other machines had to be done in the open. Two cases of the aircraft breaking adrift of their moorings were due to the moorings being at fault, however the gale was a sever one and the F.3 boats were larger than the Porte Boat for which the moorings had been designed.

Only one seaplane arrived equipped with W/T. All others had to be wired up by station staff and this took about three weeks.[26]

A Report of 23 May 1917, had noted that the *Large America* seaplanes had proved most satisfactory for submarine patrol work and large numbers would be used in Home Waters in 1918. The greater part of the establishment of 180, if not the whole output, would be needed for Home. It was considered that it was inevitable that these machines would be required for the Mediterranean, particularly in connection with the Otranto barrage.[27] As this establishment was thought to be the output available from the UK and America, it was proposed that the boats for the Mediterranean be built in Malta under Dockyard control.[28] The Air Department Progress Committee had already considered this at its Meeting on 15 May. At this Meeting it was suggested that it might be possible to have the hulls built in Malta and as the Maltese carpenters were expert boat builders, there should be no difficulty in turning out first class hulls. The wings would be built in the UK and shipped out for erection at Malta.[29] All F.2A boats were used in home waters however the F.3 was used extensively in the Mediterranean. The F.3 was to be built in Malta. The local craftsmen were skilled boat-builders and local women did the fabric work.

In November 1917 Malta reported that work on wings for F.3 boats was practically at a standstill. The next day another telegram noted that the two Rolls Royce engines "just received for the first boat" were 345-hp. They also requested a full set of F.3 drawings for the 360-hp Rolls Royce engine.[30] By January 1918 Malta reported that some of the Silver Spruce sent for main spars was defective and required 300 feet run 4 inch planks in 32 feet lengths to be sent by quickest means "to complete 12 boats." A total of 18 F.3 boats were built in Malta.

The F.3 saw service in the Mediterranean and in September 1918, three F.3 boats flew from Malta to Tripoli where they bombed a wireless station on the Gulf of Sirte suspected of communicating with U-Boats. This operation took place over three or four days. In October the type accompanied the naval attack on Durazzo in Albania.

Experimentation

An interesting experiment saw an F.2A and an F.3 (N4230) equipped with a hydrophone. The machine would alight and drop the hydrophone that was attached to a long rod on the side of the fuselage, into the water. The rod or stream-line strut according to the manual, was held to the hull by a bracket. The forward observer had headphones through which he would listen for the sounds of a submarine under water. Tests were carried out with the submarine C25 and N4230 in May 1918.

In August F.3 N4400 was undertaking trials of the Cooper servo-motor and the Cooper auto-flare night landing device "with new type stick." A Cooper servo-motor was fitted to the aileron control system of N4400. This was tried with different windmills and that of the two bladed 22 inch diameter and 3 feet 6 inch pitch "was more satisfactory." It was considered that the time taken to put the ailerons hard over should not exceed 1½ seconds.

The "Stick Night Landing Device" was to give a positive signal to the pilot of the close approach of the surface of the water in time for the pilot to flatten out. A streamlined tapered shaft was mounted on a transverse shaft in the nose of the machine. The shaft could move in and out about a foot. When the spar was out of action it lay horizontally along the fin. In calm weather the device showed promise. A landing was made with the pilot keeping his eyes off the water. The signal to the pilot was considered not as distinct as it should have been. In bumpy weather the stick developed a violent lateral sway and this interfered with the accuracy of the warning and it looked as if the stick would eventually break off the shaft.

> *Experiments are being tried with a stiffer shaft and a rounded stick instead of streamlined. A streamlined stick of larger dimensions and not so whippy can also be tried. A good deal of experiment and practice will be necessary, but it is hoped that a satisfactory method will be arrived at.*[31]

Saunders was given a contract for two experimental F.2A hulls to take standard F.2A wings. The Technical Department report for the fortnight ending 12 June 1918, reported that one hull was finished and would now be fitted with standard F.2A wings and fittings. The following report noted that the progress on the second hull was satisfactory.[32]

The F.3 was declared obsolete in September 1921, the F.2A and F.5 becoming the standard RAF flying boats.

RAF Notes on the "F.2A, V.3 (*sic*) and F.5 Boat-Seaplanes, Eagle VIII Twin Engines" record that the throttle controls were on the starboard side of the first pilot's cockpit on the F.2A and

Above: This F.2A had an experimental hydrophone fitted for detecting submarines. The operator demonstrates the device. The boom was carried along the hull and after the boat alighted it was swung out and the operator would listen for the sound of a submerged submarine. Note the way the bottom of the hull has discoloured and picked up marine organisms.

F.3, and the port on the F.5. It was noted that these "controls are not very sensitive owing to the fact that Bowden wire No.11 is used, and they have a long run."

The Engineer's cockpit was situated between the main and rear spars in rear port side of the hull. The engineer had:

(i) Distant reading thermometers for each engine.
(ii) Radiator shutters that could be controlled by the engineer to regulate the temperature of the water.
(iii) A pump for pumping extra water to the radiators. An overflow pipe from the radiators came back to the tank that held about 1½ gallons.
(iv) A Rolls-Royce dope pump with starting magnetos and "change over" switch.

The main fuel tanks were carried in the hull. The petrol system comprised a gravity tank in the upper wing that was kept full by means of a plunger type windmill pump. Petrol could be drawn from any tank by means of the pump and delivered to the gravity tank. Overflow was piped back to the main tanks. A hand plunger pump was fitted in the engineer's cockpit and used when the machine was at rest or in an emergency when the windmill pump was inoperable.

On 30 November 1918, a trial was conducted with F.5 N86 with an overload of 12,800 lbs, being 1,000 lbs in excess of the normal load.

With the wind about 8 mph and a smooth sea, a run of 57 seconds was made but the boat showed no signs of hydroplaning and continual heavy streams of spray were going through the airscrews. At the end of the run the starboard engine was vibrating so badly that it could not be run at full speed. Examining the blades of the airscrew it was found that the brass sheathing on the tips had bulged and filled with water causing the airscrew to become unbalanced. It was commented that a machine should be able to take off with at least 30% of normal load.[33] This attitude has been noted in correspondence during the war when it was recorded that the machines would always be loaded with more gear after they entered service due to the demands of that service.

Felixstowe F.5

The prototype F.5, serial N90, appeared in early 1918. It was intended as an improvement over the F.3. A typical Porte construction it embodied a number of refinements. The cockpits were open improving the pilot's backward view. The top decking of the hull was deeper while the gun positions were the same as the F.3. Aft of the wings the hull sides were fabric covered with a mahogany washboard on the lower half

Above, Below, & Facing Page, Top: The British experimented with hydrophone sonic listening devices fitted to seaplanes in order to detect submerged submarines. The Manual Hydrophones in Flying Boats and Float Seaplanes (Naval Staff, August 1918), shows the device fitted to Short floatplanes, F.2 and F.3 flying boats. N4230 was fitted with a hydrophone that was carried on the Gallows gun mount by a wooden extension to the gunbracket. The hydrophone and suspension rod is shown in the flying position along the side of the hull. The boat would land on the water and drop the device in the water to listen for submarines. This sequence of photographs shows the device and suspension rod fitted along the rear hull and in operation. The rear hatch and the side window behind the cockpit canopy are noteworthy. On N4230 the fins appear to be clear varnished while the washboard is dark (black) coloured. The wear and different tonal values of the various surfaces should be noted. Below the footstep at the cockpit is stencilled a warning not to step on the fins. Testing was carried out on 15 May 1918. (Nicolas Cooper via SEAWINGS)

Below: Close up with the device fully submerged. The operator is standing in the rear hull hatch. The dark mark in front of the serial is a chafing patch to prevent the stowed hydrophone breaking the fabric rear hull side.

Above: The hydrophone and suspension in the flying position. The hydrophone is on the end of the streamline strut.

Below: This photograph shows another experimental fitting with a hydrophone at the end of a flexible cable. Note how the hull cockade has vanished on the rear fabric covering.

Above: N4340 with the British submarine *C25* during hydrophone tests, 15 May 1918. The object of the tests would have been to determine the position of *C25*. The British and Germans experimented with seaplanes cooperating with submarines but neither came to any successful conclusions during the war. (AHT AL0655-099)

despite the problems with this arrangement on the F.2A. The F.5 hull was regarded as the best of all the Porte hulls. Bomb load was four 230-lb bombs on the underwing racks. The wing was similar to that of the F.2A and F.3 but of longer span and utilised a new aerofoil section and the horn balanced constant chord ailerons were used in place of the inversely tapered ailerons of the earlier boats. tail was essentially the same but the rudder had a balance surface. On official trials with two 350-hp Rolls-Royce Eagle VIII engines driving four-bladed airscrews the prototype, N50, had a better performance than the F.3, even in conditions of overload.

The problem of attaining and keeping aerial supremacy in the North Sea led to many suggestions as to the operation of the *Large America* flying boats as detailed in the accompanying text. The F.3 was included in these considerations. In July 1917, it was put forward by Cdr Longmore that a twin-engined fighting machine along the lines of the F.2A and F.3 was needed as an escort fighter for the formations of *Large Americas* operating in the North Sea. This new machine needed to be ready for trials when the new large calibre quick firing guns were ready for testing. The COW Automatic Gun appeared to be most promising. Porte stated that the F.5 that had just been laid down at Felixstowe could be adapted to carry two of these guns. The machine could be ready in about two months and would have a better performance than the F.3. The discussion then went on to consider whether such guns could also be used against submarines and the general opinion was that it was not possible to mount the gun such that it could perform both functions. Documentation that a COW gun was ever carried by an F.5 has not been discovered.[34]

Unfortunately the Ministry of Munitions decided on economic grounds not to introduce the new type. The Ministry did not want to supply new jigs and templates now that the production of the F.3 was well underway, and so a compromise was reached. The production F.5 used a hull similar to the prototype but with as many F.3 components as possible and was planked overall. While a very strong hull it was appreciably heavier than that of N90. The F.5 wing was virtually identical with the F.3 wing reverting to the same RAF 14 aerofoil. The F.5 did have horn balanced ailerons of constant chord. Performance of the production model with Rolls Royce Eagle VII engines was inferior to that of the prototype. The Eagle VII was used as Eagle VIII engines were in short supply but this was not the only reason that the production machines were inferior to the prototype.

Like N90 the F.5 had a balanced rudder and later horn-balanced elevators. The F.5 had a typical Porte type hull with the lower centre section integral with the hull. The upper centre section was supported by two struts, one pair placed on the centre-line of the boat. The cabane struts supporting the engines were also attached to the upper and lower centre section planes. Considerable difficulty was experienced in removing the engines once the planes were in position as it was impossible to get a direct lift on them. Air Commodore Samson noting that "replacing an engine is not an easy job without getting the boat ashore, as of course in an 'F.2.A.'

Left & Below: This H-16 N4060 was fitted with an experimental night landing stick. The device was lowered when coming down to alight and when it hit the water it levelled the boat out for a landing. The footsteps across the fins are of a different type to N4230. Note the small window above the W/T operator's position just behind the wing. This photograph of 5 September 1918, gives a good view of the engine installation. The fabric wrapped struts were probably Battleship Grey. (AHT AL917-010 & 011)

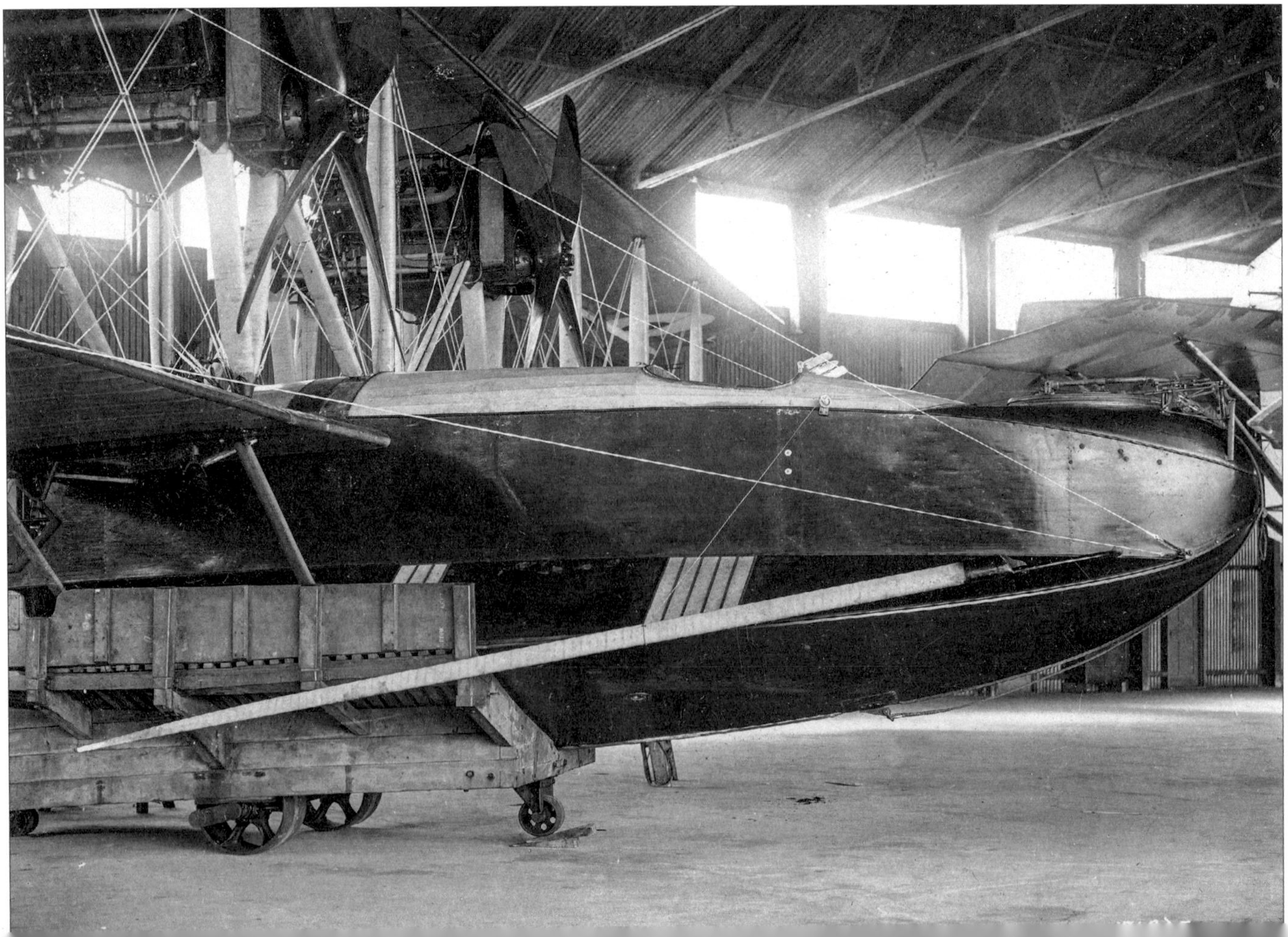

Above: N90 the prototype F.5. Note the airscrews rotate in opposite directions.

you have to remove the planes."[35] The Rigging Notes for the F.5 noted that "whenever the planes are dismantled advantage should be taken of the opportunity to overhaul the engines."

The F.5 had flown before there were any tank test results. Maj J.D. Rennie recorded that the positions of the steps on the F.5 by Porte were the result of a great deal of practical experience in takeoff and landing boats in varying weather and sea conditions. Just before the Armistice work was put in hand to fit steps to the F.5 in accordance with tank test results but the work stopped due to demobilisation of staff.

Rennie then went on to state that

the F.5 was never put into production, which was a great blunder on the part of the Production Dept, Ministry of Munitions. Instead, the F.3 wing structure, the weight considerably increased to facilitate production, and adapters fitted to take either streamline wires or stranded cables, also permanent slinging gear incorporated, was fitted to a mongrel hull, a cross between the F.5 and the F.3, and the resultant boat was called the F.5. This was done solely because the F.3 was already in production (it never should have been), and the Ministry of Munitions were against a further change as jigs, templates, etc., were already made for the F.3.[36]

In August 1918 it was noted that a further batch of F.5 drawings were received from Dick Kerr that have been traced and included with the official set for issuing to contractors. The set of drawings for this machine were now near completion.[37] It was recorded in January 1919 that the pulley bracket had been modified on as the controls were impossible to work with the present arrangement. Also that owing to the "present military situation" Liberty engines would not be installed in the F.5.[38] Too late to be used operationally in the war, the F.5, along with the F.2A, became the RAF's standard flying boat post-war until replaced by the Supermarine Southampton in August 1925.

US F-5-L

Like the F.2A/H-16 the F-5 was modified for production in the US. The dimensions were kept the same but the construction of the hull was changed and 330-hp Liberty engines were installed. Apart from an increase in size the F-5L was basically the same as the H-16 in that it had the same twin engine tractor layout, the same armament with standard 33 inch Scarff ring mounts in the front cockpit and at the aft gunner's position on top of the hull. A Gallows mount was installed inside the rear hull doors. These were referred to as Outrigger Gun Mounts in USN service. The F-5-L was to be equipped with the Davis Gun in the USN. When fitted the mount for the 9-pdr Davis Gun was located on the nose of the boat held in place by steel tube braces. The F-5-L had 22 Lewis Gun ammunition racks, two Lewis Gun parts boxes, one set of racks for Davis Gun ammunition that held three rounds and two Very's Pistol boxes. All this equipment was built into the hull.

The first NAF F-5-L was launched on Sunday 14 July 1918. On the 15th Lt Col Porte, RAF, and Lt F.E. McGill, RAF, flew

Above: N90 introduced a balanced rudder and ailerons as standard.

Above: N90 displays the fin cut-out for the rudder balance that is a distinguishing feature of the F.5. The fabric panel appears to have been painted (white) as the colour is carried onto the beading at the intersection with the ply sections of the hull.

Facing Page, Bottom: Detail of the wing overhang on N90 showing the balanced aileron and the way the cockade was carried on the lower surface of the upper wing.

Below: N4038 on the water. What appears to be the letter "E" on the hull side is more probably the door to the waist gun positions. This was the first machine from an order for 12 placed with Short Brothers Ltd. It first flew at Rochester with J.L. Parker at the controls on 10 July 1919. It served with the MAEE at Grain until at least June 1923. Final fate unknown.

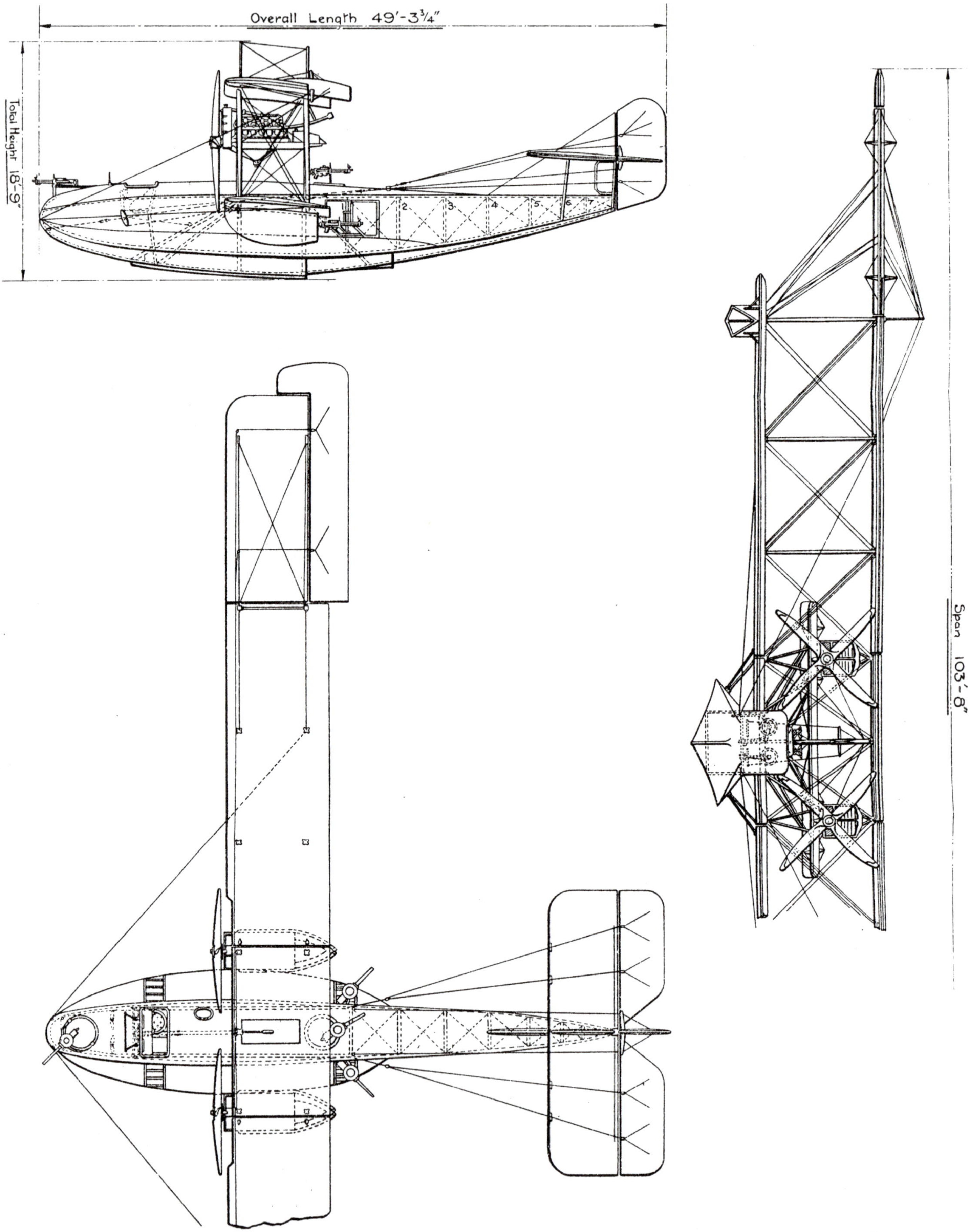
Overall Length 49'-3¾"
Total Height 18'-9"
Span 103'-8"

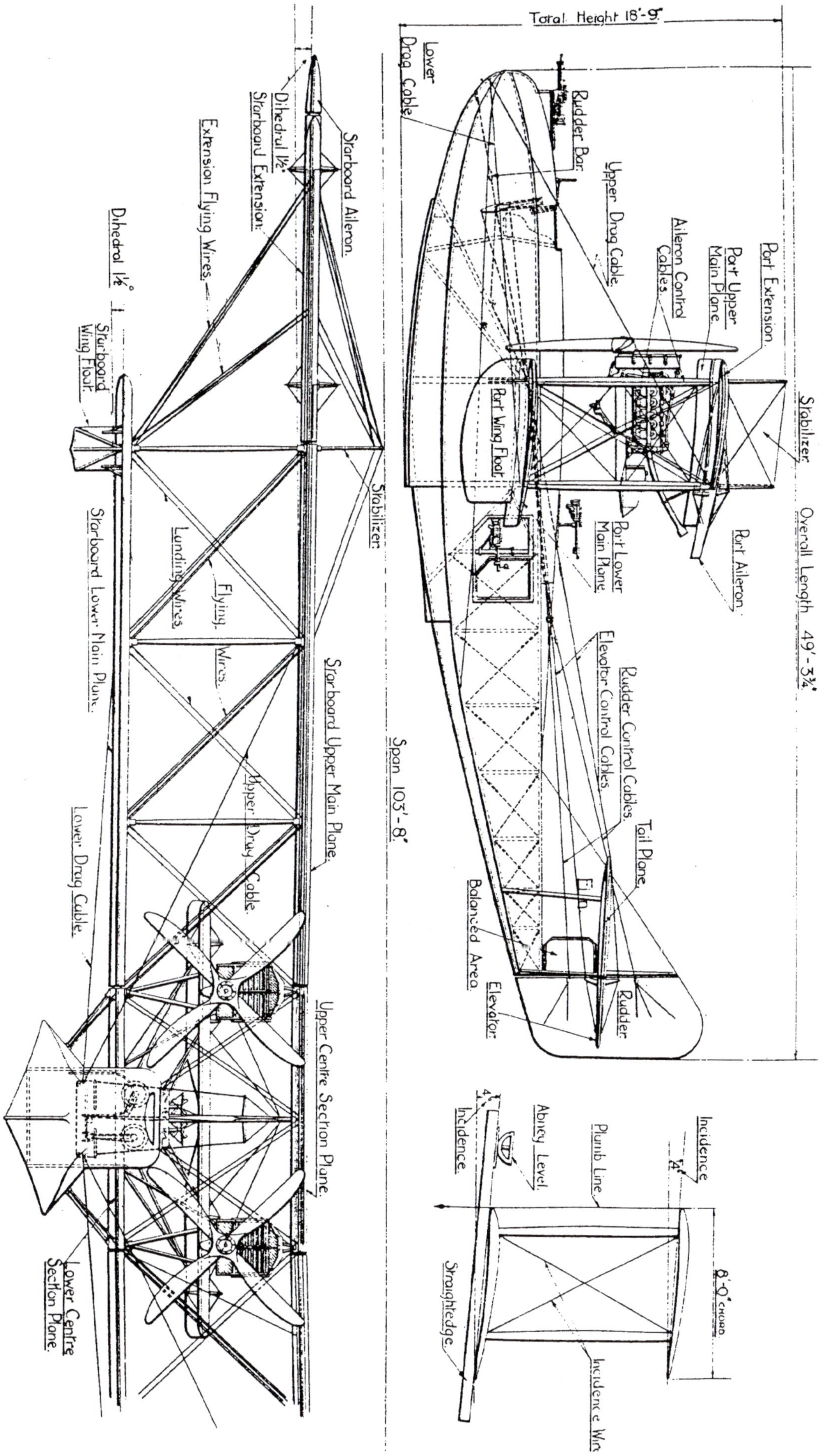

F. 5. BOAT SEAPLANE.
2-ROLLS-ROYCE EAGLE VIII.

FIG. 90.

Above: Short built F.5 N4041 (S629) appears to have a dark coloured hull overall. N4041 served with No. 232 Squadron, Felixstowe, around 1919. In April 1923 it was stated that there were 40 F.2A and 35 F.5 flyng b oats in service and as the F.5 was the later type and the establishment was 25 boats it was proposed to declare the F.2A obsolete and go on with the F.5 until a suitable replacement was available. N4041 was at Calshot in 1924 and in service until at least May 1925.

Above: N4198 on its beaching trolley with personnel. Note that the serial is white outlined on the hull. Fabric surfaces appear to be V.84 aluminium.

Above: Gosport Aviation Co Ltd built N4637 in a dark colour scheme. The cockade/roundel is now carried under the lower wing. This machine appears to have balanced elevators. This was an unlucky boat. In October 1921 while with the Seaplane Training Squadron the starboard airscrew came off forcing the boat to alight in the English Channel. On 25 April 1923, while with No.480 Flight, the rudder bar came adrift in the air and the hull stove in the following emergency alighting at Calshot. The Gosport Co was to build Porte's civil post-war designs and had the experience from its wartime contracts to do this; however, Porte's death saw the venture come to an end without any aircraft being constructed.

Right & Below: The S.536 under the serial of this F.5 indicated that N4838 was a Short Brothers constructed flying boat. It is thought that the device on the aileron is the Avro aileron balance that was tested in 1920 on this machine. N4838 was in service from 1920 to 1925.

Above: This F.5 is thought to be N4490 *Aquila* on a flight from Malta in the early 1920s. Note the cockpit layout with the crewman behind the pilot's cockpit. The different tonal values of the machine in direct sunlight should be noted.

Facing Page, Top: Saunders N4460 appears to have followed the same finishing scheme as its F.2A boats with the dark vertical stripes to the forward hull. N4460 served at Cattewater with No. 238 Squadron by September 1920. With Flt Lt G.E. Livock in command it conducted an emergency alighting on 17 January 1921, after an engine failure. The hull was badly damaged and the crew was picked up by the trawler *Verity*. (Nicolas Cooper via SEAWINGS)

Facing Page, Bottom: No details are known of this F.5 accident. The difference between the colour of the fins and hull bottom and the rest of the boat is very noticeable. These boats appear to have had the hull doped/painted white.

Above: A pristine Short Brothers constructed N4048 in post-war markings. The fabric areas appear to be white rather than V.84 aluminium. The hull appears to be black with white trim. The cockade/roundel is carried under the lower wing. N4048 made its first flight at Rochester on 27 June 1919. It was at Calshot from 1924 to 1926.

Below: N4049 at Calshot. This was the last F.5 of a batch of 12 from Short Brothers. It was based at Calshot in June 1924 and was written off after a forced alighting due to engine failure in the English Channel on 4 November 1924. Note the white outlined serial style on this batch.

F.5 Specifications

Source	1	2. Prototype Normal Load	2. Prototype Overload	4. Prototype	3. Light Load	3. Medium Load	5
Span Upper	103 ft 8 in						103 ft 8 in
Span Lower	74 ft 2 in						
Length Overall	49 ft 3¾ in						49 ft 4 in
Length Hull	45 ft 0 in						
Height	18 ft 9 in						18 ft 9 in
Chord	8 ft 0 in						
Wing Incidence	4°						4°
Dihedral	1½°	2°					1½°
Stagger	Nil						
Tailplane Incidence	2½°						
Wing Area, sq.ft.		1,409	1,409	1,409			
Weights in Pounds							
Empty	8,023	8,023	8,023	8,023	9,100	9,100	
Fuel & oil		2,097	3,121	688	694	2,862	
Military load		1,428	1,428		Nil	Nil	
Miscellaneous				199			
Crew		720	720	720	720	720	
Gross		12,268	13,306	9,630	10,514	12,682	
Speed in mph							
at 2,000 ft						88	
at 6,500 ft					86.5	86	
at 10,000 ft					78.5		
Climb in min – sec							
to 2,000 ft				2 – 45	4	6 – 35	
to 6,500 ft					16 – 30	30	
to 10,000 ft				17 – 50	38		
to 15,000 ft				35 – 20			
Ceiling				17,400 ft			
Main fuel tank	381 gal.						
Gravity tank	26 gal.						
Endurance		7 hrs	7 hrs				
Date Trial Report		05/18	05/18		05.1919	05.1919	
No. Trial Report					NM248	NM248A	

Sources:
1. F.5 Rigging Notes.
2. Chart of "Seaplanes and Ship Aeroplanes," Flight, 22.04.1920. P.441.
3. J.M. Bruce data.
4. Preliminary Report 219. "Comparison of Phoenix P.5 to Felixstowe F.5". TNA AB286/154.
5. *A Manual of Rigging for Aircraft*, Air Publication 1107, Air Ministry, UK 1925.

This Page & Following 5 Pages: This bow section from a F.5 has been restored and displayed at the East Anglias Aviation Heritage Centre of the Norfolk and Suffolk Aviation Museum. The bow section had been a potting shed in a local garden for over 60 years when it was discovered and obtained by the museum. Restored by museum members Ken Collison and Derek Small it is the largest original Felixstowe surviving artefact in the UK.

H
fly with
336

No Admittance

Above: This section of an original Felixstowe hull shows the way the planks were laid at an angle. The next layer ran in the opposite direction with a glued layer of fabric between the two layers forming a strong water tight hull when carried out correctly.

Above & Below: These two photographs have been identified as F.2A boats; however, they appear to be the same machine and the serial appears to be N90, the first F.5, although not enough of the serial can be viewed to confirm this. The finish of the boat certainly mirrors that of N90. Note the stowage for the small wind driven generator. The airscrew blades are painted Navy Grey.

Comparison of Hull Dimensions H-16 & F-5

	H-16	F-5*
Length Hull	42 ft 3 in	45 ft 0 in
Width Hull	10 ft 0 in	10 ft 0 in
Depth Hull	7 ft 7½ in	6 ft 11 in

Source: NARA RG42 File RNAF-Z-1 Vol.II.
* Since this document arose from the NAF, it is considered that it is the F-5-L hull that is the subject.

F.5 Serial Allocation				
Serials	**Number**	**Contract**	**Manufacturer**	**Notes**
N90	1		Felixstowe NAS.	Prototype.
N128				F-5L ordered from USA post-war.
N177	1		Short Bros Ltd	Metal hull with F.5 aerostructure.*
N178	1		SE Saunders Ltd.	Hollow hull with F.5 aerostructure*.
N4038–N4049	12	AS32421	Short Bros Ltd.	
N4112–N4149	39	38a/1090/C & AS4499/18 (BR347)	Dick Kerr & Co	Original contract for 18 F.3 then 32 F.5, but F.3 to N4111 only. N4125–N4149 (10), cancelled?
N4192–N4229	39	AS4496/18 (BR348)	The Phoenix Dynamo Manufacturing Co Ltd.	Original Contract for 11 F.3 and 39 F.5. N4201–N4229 cancelled, but 10 superstructures completed & delivered.
N4580–N4629	50	38a/550/C563 & AS224911/18 (BR590)	SE Saunders Ltd.	N4580–N4589 no evidence. N4590–N4629 cancelled post-Armistice.
N4630–N4679	50	38a/552/C565 & AS24910/18.	Gosport Aviation Co Ltd.	N4640–N4679 Cancelled. N4634 became G-EAIK.
N4680–N4729	50	38a/599/C627 & A26345/18 (BR620)	May, Harding & May.	Cancelled December 1918.
N4730–N4779	50	38a/604/C633 & AS26344/18 (BR621)	Dick Kerr & Co.	To have Liberty engines. Cancelled December 1918.
N4780–N4829	50	38a/598/C628 & AS26343/18 (BR622)	Phoenix Dynamo Co.	Cancelled December 1918.
N4830–N4879	50	38a/600/C629 & AS26368/18 (BR623)	Short Bros Ltd.	N4880–N4879. Cancelled post Armistice.

* Robertson defines N177 and N178 as a basic F-5

the machine. Despite developing motor trouble after a short flight and the boat being hauled out of the water to receive a new motor, Porte reported that the perfectly balanced controls operated to his satisfaction.[39]

Canadian Airplane Company (CAC) obtained a contract to build the F-5-L. Their chief engineer, Ericson, went to UK to study the plans of the machine that was referred to in correspondence as a "bomber".

Due to the problem of obtaining plans the question arose of CAC making up its own wings. Mr Ward of the NAF stated he would prefer that the wings were made according to the NAF drawings and not their own design.

The question of omitting ribs to the hull also came up – all experience at Felixstowe tended to show the necessity of transverse bracing in the hull. What was wanted was resiliency with lightness "and we obtain this with the use of ribs. When ribs are left out, the boat becomes too rigid, and in handling on trucks, the rigid bottom will fail. All experience at Felixstowe tends to show this."[40] All the CAC F-5-L boats were delivered and were apparently the same as the US produced examples.

The Atlantic and Pacific Fleets operated the F-5-L as Airboats. They were to work with the fleet in reconnaissance, range the guns and record the fall of shot and develop communication from the air to ships. They were to overcome problems that would arise with respect to these matters and establish procedures for adoption by the Navy.

These flying boats were to receive colourful schemes to enable them to be seen and identified in flight and on the water in the case of a forced alighting.

Pensacola operated the F-5-L for training. The Bureau policy was to replace the H-16 when they came up for overhaul with F-5-L boats. On 18 January 1922, Pensacola reported that it had on the station:

- 9 F-5-L (5 for use) and 7 H-16 boats.
- With 2 F-5-L under repair and 2 F-5-L awaiting repair.

The station wanted 12 F-5-L boats and in the interim asked permission to assemble for flying the following H-16 boats: A-857, A-859 and A-860 that were on the station, uncrated and stored as no F-5-L boats were available.

On 19 July 1924 Pensacola reported that its complement of F-5-L boats was:

Active	5
Being assembled after major overhaul	7
Undergoing repairs	7

Above: Construction of a lower wing for an F-5-L at the Navy Aircraft Factory (NAF). The two men are laying a straight edge over the wing and checking the tolerances. Note the smooth finish of the covered wing in the background. Rib tapes are not visible.

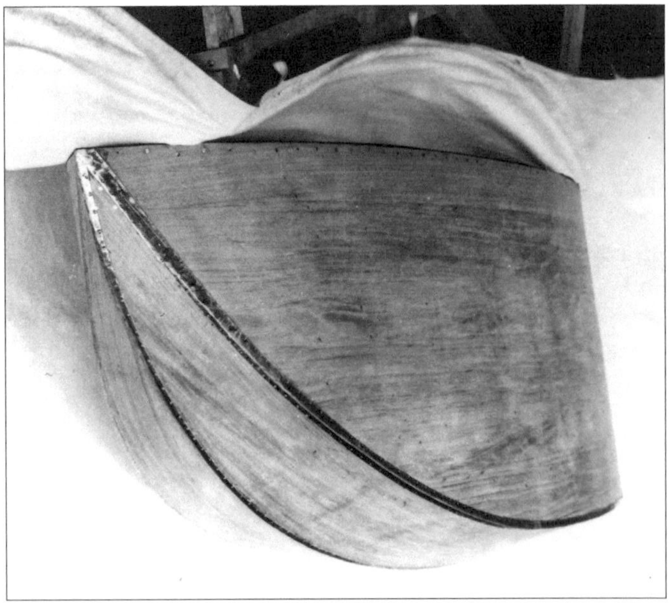

Above: A completed wing-tip float.

Left: Constructing a wing-tip float for an F-5-L.

Above: Uncovered upper wing panels at the NAF. Note the large crate being constructed in the background.

Below: Construction of F-5-L hulls; progressing from right to left the stringers are added then the planking to the bottom of the hull.

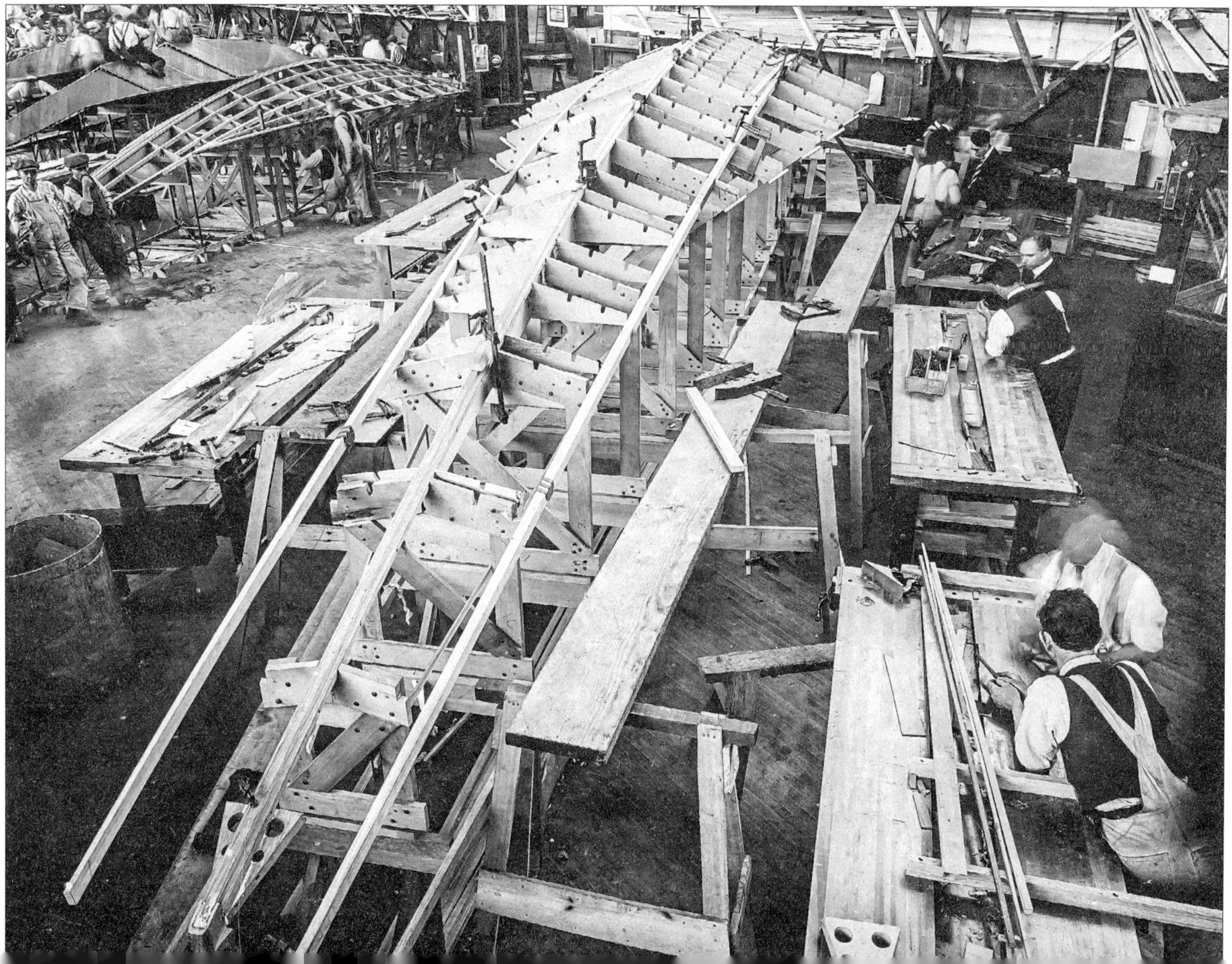

Above: F-5-L hull construction at the NAF.

Facing Page: F-5-L hull construction at the Curtiss Aeroplane and Motor Corp, 17 October 1918.

NO. 1774
10-17-18 BY E.F.
C.A.&M.CORP.
F 5 L BOAT HULL

Above & Below: Views of the planking and ply sides to an F-5-L hull at the Curtiss Aeroplane and Motor Co, 27 August 1918. The hull was constructed upside down. Note the US flag suspended in the factory, a common practice at the time. Also the large jigs and clutter in these early factories is very evident here.

Above: The Curtiss Aeroplane and Motor Corp's F5L Boat Hull Department, 31 October 1918. Twelve hulls in various stages of completion may be seen.

Below: View of the cockpit in an F-5-L hull under construction showing the left hand pilot's seat and controls.

Above: Tail surfaces added to the F-5-L hull.

Below: View inside an F-5-L hull under construction showing the complex structure of the centre hull at the wing roots.

Right: Close-up of nose of F-5-L under construction at the Curtiss Aeroplane & motor Co, 31 October 1918. The background has been eliminated on the original negative to show the forward section to advantage.

Below: An F-5-L hull packed for shipment.

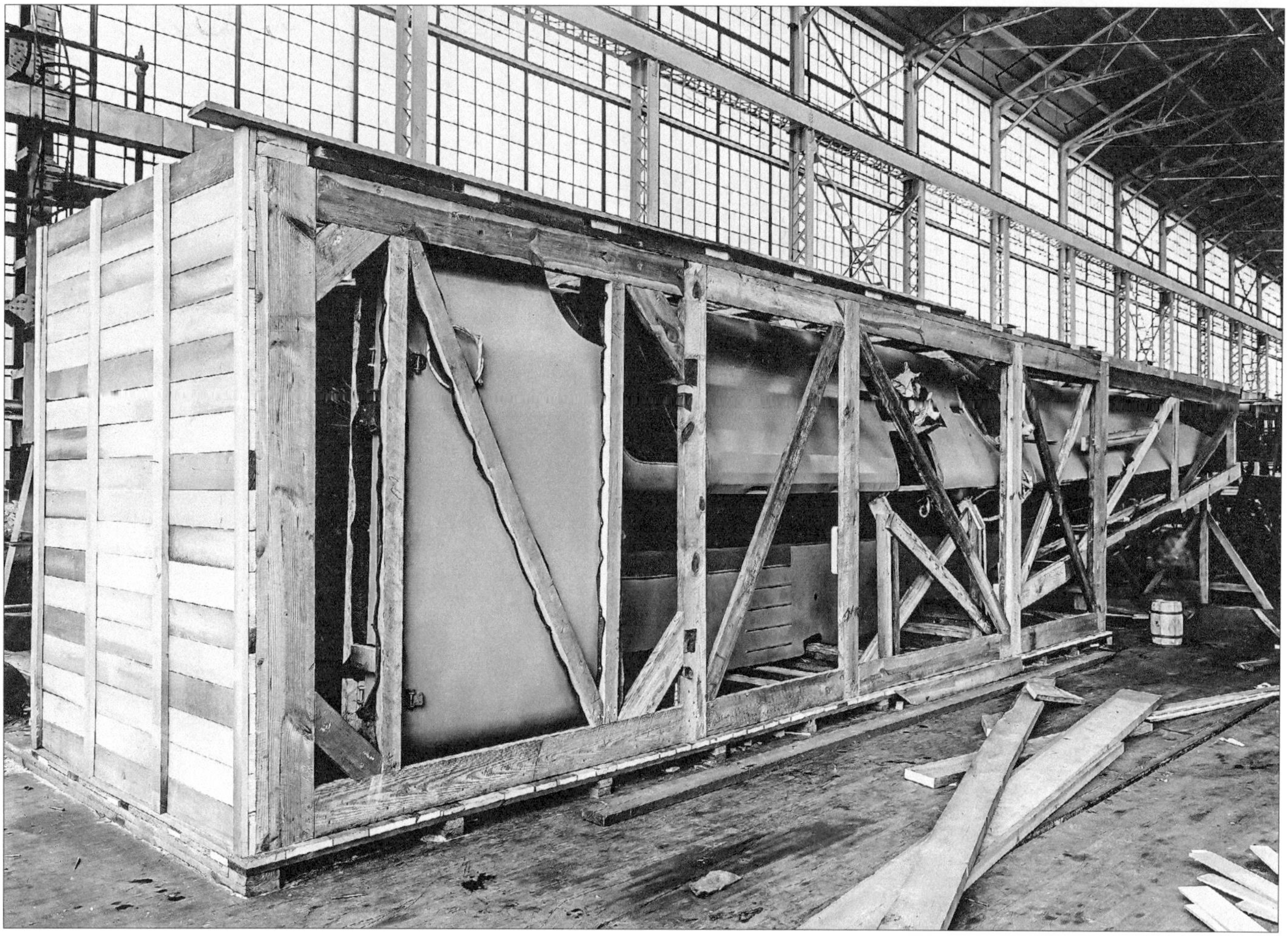

Above: The rear armament on the F-5-L. **Facing Page:** Lewis guns on the Gallows outrigger mounts on an F-5-L.

Awaiting repairs 8

USN F-5-L A-3589 made an endurance record in April 1925 of 20 hrs 10 min while covering an estimated 1,250 miles in a closed circuit off Hampton Roads under Lt Cmdr Grow as pilot.

The F-5-L was found to be excellent in the rain and sun of the tropics but their weak point was soaking. USN estimated that up to 1,000 lbs of water would be soaked into the hull. In the Pacific Fleet they were used as Liberty Boats carrying 12 to 14 men on leave parties and mail in 1919, carrying a total 1,500 pasengers and 100 sacks of mail.[41]

An aircraft mechanic recorded that he "liked flying in the F-5-L flying boats better than any plane I ever flew in up until 1940...Flying in the F-5-L was sensational flying; as the saying goes; you were flying like a bird."[42]

Experiments

There were many experiments carried out on the *America* boats and some are referred to under the various type histories. One proposal was for the fitting of folding wings to the *Large America* boats. Handley Page had experience in folding wings on large aircraft such as the O/100, and had been approached to make folding wings for the flying boats to allow more hangar space. In September 1917, it was noted that difficulties had been found owing to one of the interplane struts fouling the tail when folded. Handley Page "could only get the folding dimensions down to 43 feet. If he could be allowed to space the engines out to 18 feet instead of the present 10 feet, he could reduce the folded dimensions to 35 feet. Although the spacing of 43 feet was considered adequate as it would allow for the machines to be hangared in a Bessoneau hangar, this project never came to a satisfactory conclusion. The company had received further orders for their bombers and were devoting their energies towards this instead of getting on with the seaplane work that they already had in hand.[43]

Another interesting experiment concerned a set of Le Vaillant Speaking Tubes that were fitted to F.3 N4409. Several flights of from 3½ to 4½ hours were carried out. While it was found that a conversation could be easily carried out even with the engines running full out, there were disadvantages. The weight of the observer's tubes caused a strain on his neck as a considerable length of tube had to be allowed for his movements around the cockpit. The gun ring could not be operated as the tubes caught up on various objects. "The apparatus would be of no use during an action." The pilot was not inconvenienced to any extent with the tubes permanently attached to his helmet as the tubes were capable of being supported and he did not have to move about. They were not adopted for service us.[44]

Around May 1919, a production F.5 made an endurance flight of 14 hours 8 minutes from Felixstowe. The crew was Capt Scott, as pilot, Capt Dickey, as navigator and two mechanics. No attempt was made to alight until the Eagle VIII engines stopped from lack of fuel. The total weight of the machine with crew was 13,710 lbs of which 150 lbs was water shipped in the choppy sea before takeoff. The average speed was 55 knots and wireless communication with Felixstowe was maintained throughout the flight.[45]

Post-war the F.5 took part in many experiments. In early 1921 N4040 was fitted with a modified rudder with top balance, and tested against N4838 with the standard rudder to ascertain whether the standard F.5 rudder "was overbalanced and liable to take charge." N4040 was flown by Flt Lt E.S. Goodwin with Wing Cmdr H.R. Busteed and Flt Lt D.F. Lucking as observers. Unfortunately trials were not completed owing to engine trouble. Later flights were no more successful due to bumpy conditions. The lack of instrumentation to detect and record results was probably the reason that no recommendations were made.[46]

In another experiment N4839 was fitted with two 450-hp Napier Lion engines. In Report MN271 of 16 August 1922, the test pilot, Flt Lt C.B. Dalison, AFC, recorded that the

> *seaplane is very nose heavy on climb but trims at nearly full throttle flying level low down at about 78 knots and 2000 r.p.m., under which conditions it flies "hands off". The above with the maker's tail setting.*

To the rest of remarks regarding Stability and, Controllability and Manoeuvrability, the Report notes all "As Eagle 8.F.5." No bombs could be carried externally due to position of docking chocks. It was considered that the poor performance was partly due to the excessive radiator size and the Vickers Vernon airscrews used for the test. "The engines are overloaded by the Vickers Vernon propellers." It was considered that the "improvement in performance over that of the standard Eagle 8, F.5. is so far inconsiderable that no useful purpose would appear to be served by the present conversion scheme."[47]

There were two special F-boats. N178, that featured a deep Saunders hollow-bottom hull. This machine has standard F.5 wings rigged with a slight stagger. N177 followed with an all metal hull by Short Brothers and known by that firm as their S.2. The hull was an early metal monocoque and proved its strength when the machine was stalled into rough water from about 30 feet and remained undamaged and watertight.

Post-War

The British were anxious to catch as much of the post-war civil and military market as they could. Their main rival was France as the Germany aviation industry was hamstrung by the Allies. Two F.5 boats, N4041 and N4044, set out from Felixstowe on 11 July 1919. N4044 made a round Scandinavia flight of 2,450 sea miles in a flying time of 40 hours 40 minutes as a "demonstration of the commercial uses of flying-boats." The flight was noteworthy that in the 27 days no kind of trouble was experienced and the low-compression Rolls-Royce Eagle VIII engines 'worked magnificently throughout." The flight took in Norway, Denmark and Sweden.[48]

The ELTA Exhibition at Amsterdam in 1919 was to be a showpiece for aviation, especially civil aviation. Aircraft companies from all over Europe presented their products at the

F-5-L Specifications				
Source	**1**	**2**	**3**	**4**
Span Upper		103 ft 9¼ in	103 ft 9¼ in	103 ft 9¼ in
Span Lower		74 ft 3⅞ in	74 ft 4 in	74 ft 3⅞ in
Length Overall	49 ft 3¹¹/16 in	49 ft 3¹¹/16 in	49 ft 3¹¹/16 in	49 ft 3¹¹/16 in
Length Hull	45 ft 3 in	45 ft 3 in	45 ft 3 in	
Height	18 ft 9¼ in	18 ft 9¼ in	18 ft 9¼ in	18 ft 9¼ in
Chord			8 ft	8 ft
Gap			8 ft 10½ in	8 ft 10½ in
Dia Prop	10 ft 6 in			
Incidence	4°	4°	3° 40'	4°
Dihedral (Lower)		1½°	1° 30'	1° 30'
Aerofoil				Mod RAF6
Areas in Sq. Ft.				
Wing upper			848	
Wing lower			546	
Wings total	1,397		1,394	1,278
Ailerons (2)			118	119
Horizontal Stabiliser				120.9
Fin				34.7
Rudder				33.5
Elevators (2)				55.3
Nonskid fins (2)				33.1
Weight in Lbs.				
Wing structure				2,613
Hull				2,331
Engines, etc				3,232
Electrical				64
Repair				107
Navigation				48
Medical				10
Radio				260
Ordnance				1,190
Crew (4)				720
Fuel & Oil				2,425
Useful load			5.224	
Military load			1,405	
Total				13,000
Speed			57–90	
Climb to ft			2,600/10 min	

Source: 1. *Characteristics of a Boat Type Seaplane During Take-Off*, Appendix taken from Bureau of Aeronautics USN Performance Chart. Sketch shows F-5-L with final modifications.
2. NAF three-view.
3. "U.S.A. Navy F-5-L Flying Boat", *Flight*, 31 July 1919.
4. *F-5L Flying Boat Handbook*.

exhibition and passenger flights introduced the new method of transportation to the public.

Gosport built F.5 N4634 flown by Lt-Col Ralph Hope Vere was the first F-boat to attend the exhibition. Banned from flying by the British Air Ministry N4634 was only available to invited guests and was not available to the general public. To highlight the British presence Sir Fredrick Sykes, the Controller General of Civil Aviation, arrived with five RAF F.5 flying boats at the naval air station of Schellingwoude on 12 August. These flying boats returned to Britain the next day. F.2A boats N4439 and N4441 were flown to Holland for the ELTA exhibition with N4441 also giving demonstration flights. These two F.2A boats were presented as Vickers exhibits as S.E. Saunders had been taken over by Vickers.

Canada

Canada received eleven F.3 flying boats as part of the Imperial Gift of 1919.[49] They were given civil registrations G-CYBT (N4016); G-CYDH (ex-N4009); G-CYDI (ex-N4010); G-CYDJ (ex-N4011); G-CYDQ (ex-N4014); G-CYEN (ex-N4015); G-CYEO (ex-N4181); and also N4012, N4013, N4178 and N4179 that did not receive civil registrations but were held as spares.[50] A single Curtiss H-16 N4905 (G-CYEP) was included in the Gift but two ended up in Canada. The second was N4902 that was not registered and apparently used for spares.[51] The US Government also gave the Canadians a number of Curtiss HS-2L flying boats that had operated from Nova Scotia during the war.

The flying boats were used for summer patrols over areas where forest fires could break out. By 1922 all the Canadian aircraft were worn out and had not been reconditioned. The last F.3 was withdrawn in 1923, their replacement was the Vickers Vedette that was designed for Canadian conditions and built in Canada by Canadian Vickers at Montreal.

Chile

The Chilean Navy acquired an assortment of British aircraft after the war as payment for the Chilean ships that were under construction in Britain at the declaration of war and were taken over by the British. Sopwith Baby and Short 184 seaplanes were received followed in 1920 by a single F.2A (ex-N4567). This boat had open cockpits and balanced ailerons. Commissioned on 17 October 1921, when it was christened with the name "Guardiamarina Zañartu" after the first naval pilot. It was replaced by the Dornier Wal in 1926 and made its last flight on 15 February 1928.

Japan

The Master of Sempill's British Naval Air Mission to Japan saw British aircraft purchased for use by the Japanese Navy. Before their arrival in April 1921, a small number of Avros and F.5 boats were ordered by Japan. The ten F.5 boats came from Short Brothers. The first to be erected was launched and flown on 30 August 1921, by John Parker. The F.5 "proved a valuable asset to the work of the Mission" and a number of long distance flights were made by trainees. It was the practice to finish each course with a cruise around the shores of Japan. The distance was usually over 1,500 miles and these took more than 9 hours to complete in stages, the longest stage usually being about 500 miles.[52]

The F.5 was to be built in Japan, Short Brothers sending 20 personnel under engineer Dodds to Japan, the contingent arriving in April 1921. They worked at the Yokosuka Arsenal's Ordnance Department. The Arsenal erected the imported F.5 boats with the first taking to the water in April 1921. British sources state that in 1922 three Short-built F.5 boats with Napier Lion engines were obtained from Britain.[53]

An US intelligence report noted in March 1923 that two F-5 flying boats had just been completed at Yokosuka for the Naval Air Service at Sasebo. The aircraft were serialled 51 and 57.[54] The first F.5 constructed entirely of local materials was built at the Hiro Naval Arsenal with assistance of Short Brothers personnel. Later some 50 F.5 boats were built by the Aichi Tokei Denki Co of Nagoya. In 1925 the Hiro Arsenal powered an F.5 with licence-built 400-hp Lorraine engines, followed by another with 450-hp Lorraines. Designated F.1 and F.2, they were not selected for production as it was realised that the F.5 was outdated. Ten were built by the Hiro Naval Arsenal (Hiro Kaigun Kosho), ten by the Yokosuka Arsenal and forty by Aichi. Several impressive long-distance flights were made by the Japanese F.5 boats, some over nine hours. Some survived still in service in 1929. A report from the Naval Attaché, Tokyo, on 18 January 1929, stated that 16 F-5 boats were in first line service with 6 in reserve.[55] Although the type had a good history of service there were also many accidents mainly due to improper maintenance, engine problems and bad weather. Japan was to go on and build some of the best flying boats in the world culminating in the mighty Kawanishi H8K "Emily" of WWII fame.

Australia

Australia sought to obtain Short F.5 flying boats and the serial sequence A11 was allocated to the proposed boats but financial concerns led to the purchase being abandoned. Two wooden hulled Supermarine Southamptons were eventually obtained and they took up the A11 serial prefix.

Netherlands

As related in the text N4551 fell into the Netherland's hands. A number of *Large America* flying boats ended up in the Netherlands commencing with H-12 8693. This boat was delivered to Felixstowe on 3 June 1917, and was forced down due to engine failure on the 24th the following October. The Dutch torpedo boat *G15* found the boat anchored in the Deurloo. The crew of Flt Lt H.C. Gooch, LM C.W. Sivyer and 2AM B.M. Millichamp were taken on board the torpedo boat whereupon the flying boat sank. Sources vary as to the fate of this machine with British sources saying the crew sank the boat, which appears to be most likely, while there is a suggestion in Dutch sources that there may have been a partial salvage of the machine.

F-5-L Serial Allocation				
Bureau No.	**Number**	**Contract**	**Manufacturer**	**Notes**
A-3333 to A-3382	50	37211 CR493; SE818; Bu18	CAL	A-3363 to A-3382 cancelled.
A-3559 to A-4035	480	RNAF-Z-5	NAF	343 cancelled at random: A3618 to A3658 A-3684 to A-3782 A-3801 to A-3858 A-3881 A-3883 to A-3835 A-3941 to A-4008 A-4014 to A-4035
A-4036 & A-4037	2	RNAF-Z-5	NAF	F-6. NAF modification of F-5-L.
A-4038	1	RNAF-Z-5	NAF	
A-4281 to A-4340	50	N3409 CR70; SE154; Bu19	Curtiss A & M	
A-4470 to A-4819	350	N3935 CR99; SE236; Bu19	Curtiss	
A-5259 to A-5458	200	RNAF-1	NAF	Cancelled.
A-6557 to A-6559	2		Hampton Roads NAS	Erected from spares & salvage.

Below: F-5-L hulls at the Curtiss Aeroplane & Motor Corp. As the hulls progress down the line the tail surfaces are added. Curtiss Photo No.1874.

Above: A colourful F-5-L of the Atlantic Fleet Scouting Squadron coming up to the stern of the USS *Shawmut* seaplane tender at Guantanamo, Cuba, February 1921. The dark colour to the upper wing is thought to be chrome yellow. The Atlantic Fleet boats were given colourful hulls for the same reason that dazzle painting was adopted by the British in late 1918.

Below: An F-5-L on a seaplane barge (lighter), Guantanamo Bay, Cuba, circa 1921. This machine has a colourful hull and the large fin post-war modification. Note the two men on the upper wing and the trousers hanging out to dry under the starboard wing.

Above: "In the enemy's smoke." This F-5-L is flying above a smoke screen laid down by the warships during the June 1922 Fleet manoeuvres. The hull is brightly painted and the fabric surfaces appear to be aluminium doped. The top surface of the upper wing is chrome yellow.

Below: The F-5-L was used for parachute training as shown here (probably early 1920s). The fish shape is the weight for the wireless trailing aerial.

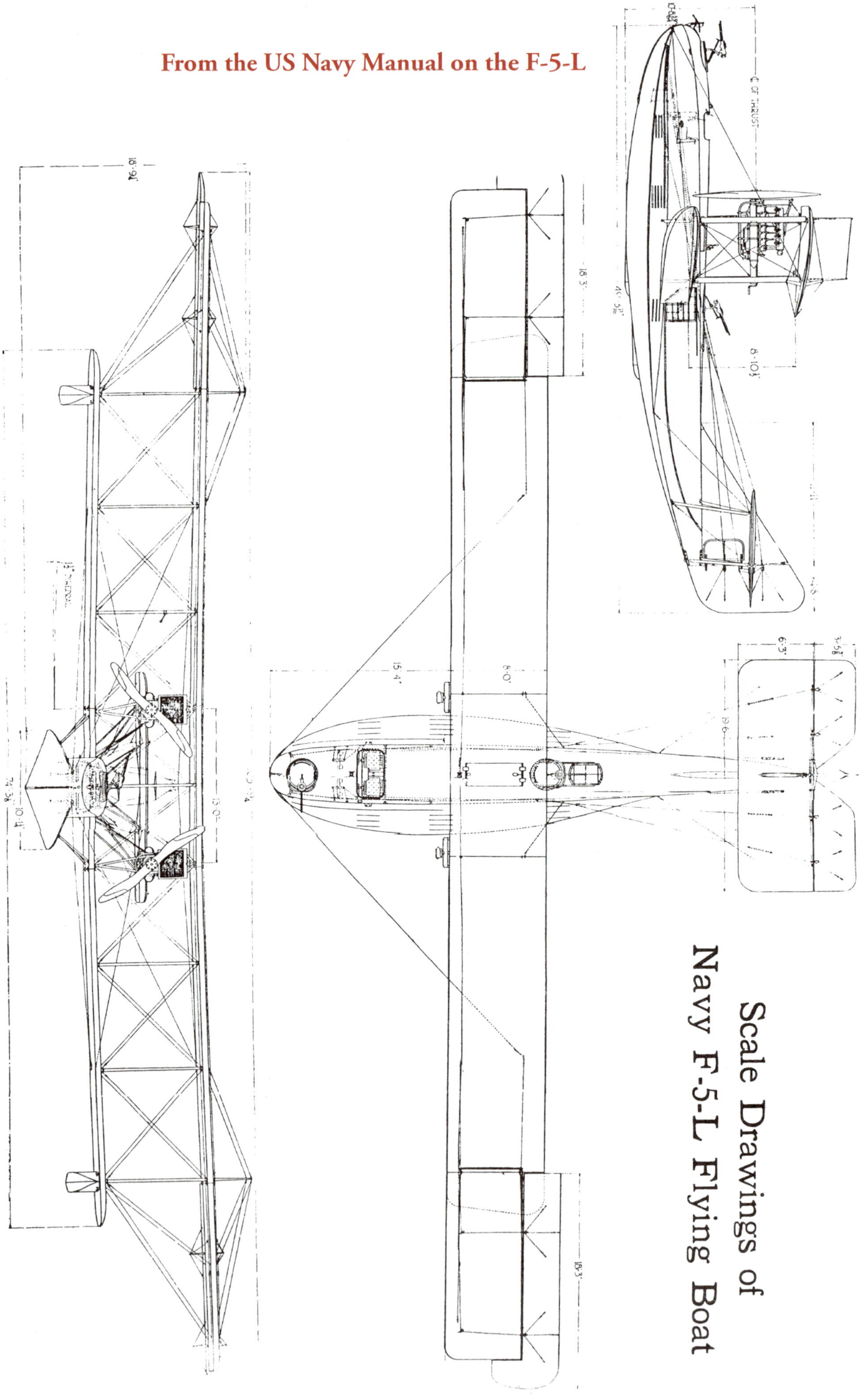

Scale Drawings of Navy F-5-L Flying Boat

F-5-L A3859 assigned to NAS Guantanamo Bay in March 1920; it was deleted in August 1922. It was one of five F-5-L aircraft, the others being A3601, A3603, A3662, and A3665, that undertook a six-week cruise through the Caribbean in support of the fleet.

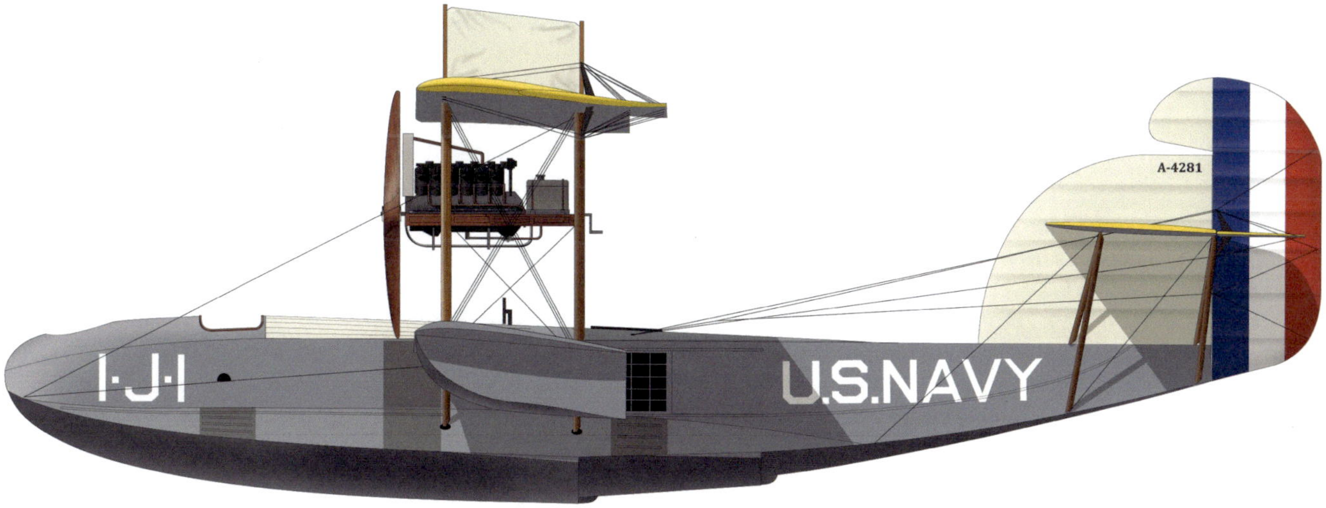

F-5-L A4281 assigned to Utility Sqdn 1 at NAS San Diego, 1926–1928.

F-5-L assigned to Patrol Sqdn 14, USN.

Above: A-3859 powering to get onto the step for takeoff. The insignia on the bow is of a flying fish. In this photograph the machine appears to be in navy Grey overall. The aircraft number (the last two numbers of the Bureau No. is not present on the bow but carried under the hull. This machine appears to have the WWI style national insignia. Note the black walkways on the fins.

Below: The second photograph (below) shows the machine in Guantanamo Bay, Cuba, with what appears to be a light (white) colour forward hull with the flying fish insignia in the oval. The wings may be aluminium doped. The horizontal tailplane upper surfaces are dark (chrome yellow) coloured. The cockpits and engines all have canvas covers.

Constructed at the NAF, A-3859 was reported as complete on 1 August 1919. From 27 September to 15 October the boat was assembled and flown to the Air Detachment Atlantic Fleet. Before leaving the NAF, the machine was weighed immediately after assembly on the 12 October and before being launched. Weight was 11,350 lbs without crew, guns or ammo and no other equipment that was not a permanent fixture. On 4 November it was delivered to the USS *Shawmut*.

From 14 February to 31 March 1920, Five F-5-L boats of the Airboat Squadron, Air Detachment, Atlantic Fleet. (A-3601, A-3603, A-3662, A-3665, and A-3859) undertook a long cruise. This is recorded in detail as it gives an indication as to the work performed by the F-5-L in service with the Fleet. On the 14th the Squadron arrived St Thomas, Virgin Islands. 21st left for Ponce, Puerto Rico. 26th left for Mayques, Porto Rico with *Shawmut* and *Sandpiper*. Left 28th for Aquin Bay. All had difficulty in taking off. By starting one behind the other and using the slipstream of the airboat in front to create wind, all finally got away. 1 March, left for Kingston, Jamica. 4th left for Guantanamo, Cuba. Overtook Battleship Squadron. 5th to 15th very little flying. Aircraft given maintenance. On 16th left for Cape Cruz arriving Santa Cruz del Sur after 2 hours 45 minutes. On 17th two airboats were sent to act as escort for shipplanes flying off turrets. Nieuport 28 A-5794 suffered a forced landing and sank almost immediately. Pilot rescued by airboats. 18th two airboats repeated escort duties. Shipplane A-5798 missing. Search continued until dark. Found by A-3786 following day. 19th SantaCruz del Sur, Cuba, Commander Atlantic Fleet. Recommend remove engines for complete overhaul. 20th to 25th spent in repairs and maintenance. On 26th entire squadron proceeded to Manzanillo. A-3859 required engine replacement. As change over not completed was towed by Sandpiper to Manzanillo, Cuba, on seaplane barge. Bomb dropping practice. MF boats accompanied squadron during this flight. 27th to 31st spent in replacing engines, bomb dropping and night landing practice, and maintenance.

On 24 June the machine was sticken in error. The boat was shipped to Hampton Roads and finally stricken on 2 August 1922.

Above: A-3601 on 30 December 1919, at Pensacola. Reported as 100% complete on 18 July 1919, A-3601 was selected as a replacement F-5-L for the Atlantic Fleet, and was assembled at the NAF and shipped to the Air Detachment, Atlantic Fleet during 27 September to 15 October 1919. In November it was recorded as operating with the USS *Shawmut*. As an Airboat, A-3601 had a colourful hull with white stripe separated by the code 01. The last two numerals of the Bureau No. were used as the aircraft codes and were usually also carried on the nose of the bow. This Airboat accompanied A-3859 on its 1920 cruise. On one occasion during the cruise a forced alighting was caused to A-3601 by the throttle wire cutting through on a sharp corner where the wire was secured by a sheave. The flying boat had to be towed in with a very heavy sea running which resulted in one wing being damaged, and it was undergoing repairs on the 11th for damage caused by this incident. On 4 August the boat was at Santa Cruz del Sur, Cuba when the camshaft lost two teeth causing the engine to be surveyed. On the 10th it was shipped from the *Shawmut* to the NAF presumably due to its condition as it was recommended to be stricken in the following December, that taking place on the 30th of the month.

Below: The upper wing of A-4281 is chrome yellow. This boat carried late markings and has all the modifications to the tail surfaces and hull. The hull markings "I-J-I" are thought to show that this machine was the first plane in Utility Squadron 1. In August 1918 reported as 30% complete, and in November as 100% complete. On 18 November it was shipped to the NAF and presumably placed in storage where it was recorded on 22 October 1920 when it was recommended for sale. On 17 April 1926, the boat was shipped to the Scouting Fleet operating with the USS *Wright*, and was received on 5 June. It was returned to the NAF in August, probably for overhaul as it was shipped on 11 September to San Diego to Utility Squadron 1, VJ-1. The boat lasted until 1 May 1928 when it was recommended that it be stricken. The hull lower edges of keelson under gas tank was discoloured and spongy, etc., due to service and age. On 27 June, with a TFT of 473.35 hours, A-4281 was stricken.

Above: An early F-5-L in flight in WWI colour scheme.

Below: The USS *Wright* was a flying boat depot ship. After launching in April 1920, the ship was converted to a lighter than an aircraft tender. The ship was commissioned at the New York Navy Yard on 16 December 1921. The ship had a balloon well built into the hull to enable a kite balloon to be carried by the ship. After discharging her kite balloon the ship acted as a seaplane tender.

Above: This F-5-L with modified fin and rudder carries the late national insignia and bears the last two numerals of its Bureau No. on the upper wing. Compare this aluminium doped wing with those painted chrome yellow. Note the black stripes at the outer interplane struts.

Below: A late F-5-L in flight with the enlarged fin and balanced rudder and external stringers to the rear hull. This machine is in post-war colours with star in circle national insignia. The illegible Bureau No. is carried on the fin.

Above: This F-5-L has all the modifications to the fin and rudder and hull. The station code is now carried on the hull and on the centre section of the chrome yellow coloured top wing. The "P" is thought to indicate "Patrol."

Below: Although it bears late markings this F-5-L has the original fin with modified balanced rudder. The code is thought to indicate that this is the 4th plane in Utility Squadron One. The meaning of the "X" added to the code is not known. Note the bulge at the bow.

Above: The codes on the upper surface of the top wing have been over-painted. This machine has the late hull modifications but early fin with balanced rudder. Note that the upper surface to the top wing and the horizontal tail surfaces have been doped chrome yellow.

Below: The purpose of the bulge at the front of this F-5-L boat's hull is unknown at this time. The aircraft number "61" is repeated under the lower port wing.

Above: Capt Mustin, early USN aviator, leaving F-5-L No.4 in San Diego, March 1921, after Panama trip. Note the well worn hull painted a light colour on its upper surfaces.

Above: With a striped hull that appears to have different colours on the hull and fin over the standard Navy Grey finish, this F-5-L has the large fin but retains the unbalanced rudder. There are two occupants in the front gunner's cockpit.

Above: A-4293 appears to be aluminium coloured overall. The fin modification is present but not the balanced rudder. The machine is being refuelled. A small tractor for moving the boats is visible under the port wing, a saving in manpower when compared with the operations in WWI when many hands were required to shift these large flying machines. A4293 was was reported as 50% complete in November 1918. It was apparently stored at Buffalo until 2 April 1919, when it was shipped to Philadelphia, presumably to NAF for overhaul before being shipped to Coco Solo in January 1920. In June 1923 it was returned to the NAF arriving in September where it was recommended it be stricken. This occurred on 30 September 1927, with a TFT of 116:30 hours.

Above: This F-5-L appears to have white stripes applied over the basic Navy Grey finish. Note the ships in the background. The dazzle painted boats were operated with the fleet.

Above: A-4308 displays the lower surface markings on a late F-5-L with the hull strengthening modifications. The hull is the same colour overall. Curtiss built A-4308 had a relatively long life. In November 1918 it was reported to be 40% complete and was therefore selected to be finished as were all the 50 machines of this order. Initially stored at Buffalo, NY, it was shipped to the NAF in March 1919, and was presumably overhauled before being shipped to the USS Shawmut for the Atlantic Fleet with Liberty engines Nos. 02772/3841 (stb); 03822/5424 (port) installed. On 30 December 1920, Shawmut reported that the bow was pushed in about 6-8 inches from a point above the waterline to the machine gun mount. The plane did not make any water and no other part of the hull was damaged. Skid fin on starboard wing was broken but the overhang was not broken. The boat was beached and the bow covered with canvas and the crew then proceeded on their journey. Four longerons and the main keel were broken; all fore and aft main ribs were broken, all to 2 feet aft of the bow. All cowling ribs were broken forward of gunner's cockpit. Both starboard and port sponson longerons were broken 2 feet aft of the bow. No mention of how this occurred is given in the report. (Another document gives the date of this accident as occurring in 1921.)

On 25 March 1921, while operating with the *Shawmut* out of at Guantanamo the boat suffered a forced alighting at sea due breaking generator drive shaft. Wings were to be repaired and the engine replaced. Shipped to the NAF in May it was reported that the hull and surfaces to have slight repairs and all surfaces to be reconditioned before it would be fit for service. It must have been considered uneconomical to repair as the boat was stricken in July but reinstated in October for use with shore stations but not with the Fleet. A reconditioned A-4308 was shipped to Pensacola on 31 June 1922. Pensacola reported that the condition of the hull and fabric was only fair, but cleanliness was very good. The boat remained at Pensacola until 1926 when it was shipped to the NAF in October and was stricken on 27 December the following year.

Below: Another F-5-L in a colourful scheme. The upper wing has bands of light colour (white/aluminium dope?) and the skid fins have what appears to be an insignia on their outer side.

Above: With canvas cover over the cockpits F-5-L A-3863 sits on the end of what appears to be a breakwater at Guam 1923. Note how the boat is supported at the wingtip floats and the rear hull. This boat was completed in early 1919 and may have been stored at the NAF as it was the subject of a maintenance report there in May 1922. It was allocated to the Atlantic Fleet Torpedo Squadron operating with the USS *Sandpiper* later that year. In 1923 it was assigned to the USS *Wright* and appears to have spent the time when it was not with the Wright at Hampton Roads. On 27 March 1924, Hampton Roads reported that the hull bottom planking needed replacing and recommended that the boat be stricken, the hull burnt and all useable material taken into store. This was carried out on 7 May 1925. The boat had a TFT of 399:40 hours at the time.

Left: USS *Sandpiper* with remains of an F-5-L boat, 13 November 1920. The dazzle markings of this flying boat can just be made out on the rear hull.

Below: This F-5-L, machine "03", has a colourful hull. Note the canvas covers on the airscrews.

Above: The top surface of the upper wing on this F-5-L has large chrome yellow patches. The aircraft code "83" is marked on the centre section but is not visible on the bow. Note the crewman standing in the front gunner's cockpit.

Below: Two F-5-L boats in flight. The closer appears to be in aluminium overall while the other, A-4329, appears to have a Navy Grey hull. Both machines have the enlarged fin and balanced rudder and carry their last numerals of their Bureau Nos. on the centre section as the aircraft's code number. A-4329 was constructed by the Curtiss Aeroplane and Motor Corp as part of a batch for 50 that was allowed to stand after the Armistice. Originally estimated to be ready in December 1918, it was shipped on 13 March 1919 to Philadelphia, probably for storage at the NAF. On 20 November 1920 it was shipped to Pensacola, and the following year to Hampton Roads. A report of 20 November 1922 condemned the hull and the seaplane was stricken on 2 January 1923.

Right: In Navy Grey overall with codes "1-S-8" on the rear hull and an interesting insignia on the forward hull, this F-5-L was photographed in Newport Harbour, Rhode Island. The "8" was repeated on the bow of the boat.

Below: This F-5-L has the strengthened hull with ply sides. Although it cannot be seen clearly there are some numerals on the forward rudder stripe. The Bureau No. was normally carried on the central white rudder stripe.

Above: A-3565 carried the individual aircraft number "2." In this colour scheme the upper wing is a light colour (white), probably the same as the hull checks. This F-5-L was constructed at the NAF. As at 30 September 1918, A-3561 to A-3577 were reported as 65-100% complete. On 17 September the boat was loaded onto a car for shipment. In 1920 it was offered for sale at the NAF. In may have been in service by 21 January 1922, at Hampton Roads when a Maintenance Report was made. On 18 December 1922, it was flown from the NAF to Hampton Roads for the Atlantic Fleet. It was probably with the Fleet when the high visibility colour scheme was applied. During 1923 to 1925 it seems to have been assigned to Hampton Roads and the USS *Wright* alternating between the two. On 1 July 1925 it was reported by Pearl Harbor as having been recovered prior to the departure of VS 1 from Hawaiian waters. A Report of 15 December stated that all the lower fittings were found to be badly corroded and unfit for further use, especially the main fittings in the tank section. As a result the boat was stricken from the list of Navy aircraft on 28 January 1926.

Below: Three flying boats and one floatplane may be made out on the water. Two of the boats are at the stern of the two ships.

Above: The station that this F-5-L came from is clearly marked on the hull along with the person who was using it for Christmas 1929. It would be interesting to know if it flew festooned as it is depicted here.

Below: The fin and rudder of this F-5-L display a dirty appearance unlike the hull. Note the access ladder leading to the side gunner's position.

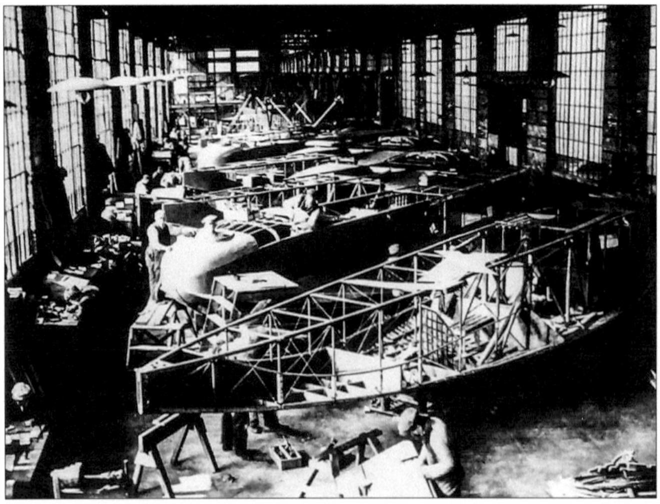

Top: The first F-5-L produced by Canadian Aeroplanes Ltd, A-3559. In July 1918 reported 100% complete and shipped to NAF for testing on the 30th for experimental purposes. Recorded at Hampton Roads on 16 November. Received at NAF on 11 February 1919. On 14 July while attached to the NAF A-3559 crashed to a total wreck. Ens Harold Livermore Roehrig and two 2 men were killed. 6 August the boat was stricken. TFT 17:30 hrs.

Left, Above Left, & Above: F-5-L boat construction at Canadian Aeroplane Ltd.

Above & Below: F-5-L flying a Rear Admiral's staff flags attached to the interplane struts and marked with two stars on the forward hull. This is Rear Admiral Moffett's aircraft. The rear hull gunner's opening has been extended. While this machine has the late hull and rudder modifications it retains the early fin.

Above: The HR on the bow of this F-5-L denotes that it was assigned to Hampton Roads.

Below: The first F-5-L built by the NAF, A-3559. The machine is Navy Grey overall. The original recessed rudder balance may be seen under the horizontal tailplane. Reported as 50% complete in July 1918 and 100% complete the following month it was ready on 11th of that month. If flew from the NAF to Hampton Roads on 28 August and was still there on 30 November when it alighted due to a heavy squall. As the wind died down the pilot tried to reach Hampton Roads before dark. Ran into heavy rain such that it was impossible to tell which direction seaplane was heading as compass could not be read. Absence of windscreens made it additional difficulty so the pilot attempted to land but could not see water due to the rain. He levelled off too high and side slipped to right. Hull began to fill and the crew took refuge on tail. Recommended be stricken, cause held to be the improper placing of the windscreens and light compass. On 17 September this was carried out. Another source indicates that this may not have been carried out immediately as it has the machine still at Hampton Roads in November. (AHT AL1171-046)

Above: F-5-L Bureau No. A-4036 being prepared for launching down the slipway. Built at the NAF in September 1918 it wa reported as 20% complete, and 90% in November. On 17 January 1919, Ensigns Stillwell and Stinson flew the boat to Hampton Roads who reported that the starboard and port propeller stockings were badly damaged; the overhang load wire on the port wing was broken; the trailing edge on both upper wings and the lower port wing was badly warped; and there was considerable water in the hull that had to be pumped out and the bottom thoroughly cleaned inside and out. It would require two week to get ready for testing. On 12 February Hampton Roads reported that the machine had an enlarged vertical fin fitted for trial flights. On 10 April it was reported as damaged and on 9 June the boat was recommended to be stricken as "damaged", this taking place on 9 June. TFT was 18.39 hours.

Below: F-5-L A-3864 with tandem engine installation. The rear engine has a four bladed propeller. Reported complete on 15 August 1919, by November A-3864 was undergoing tandem engine trials at the NAF. In June 1921 it was shipped to Hampton Roads where the trials were to continue. IT was still with the Experimental Squadron on 27 April 1922, when it was awaiting propellers from the NAF. It is assumed that it was converted back to the normal engine installation as it was recorded with Scouting Squadron No.1 assigned to the USS Wright in May to November 1923. On 27 March 1924, Hampton Roads reported that the vertical keel and a large percentage of the thwart ships frames and longitudinals had deteriorated due to constant contact with water. All bottom planking was in need of replacement. It was recommended that the hull be reconditioned and placed in reserve. Another report on November 1925 stated that the boat had been erected in 1919 and used as a station plane until 1921. The vertical keel, etc., were rotten and the boat was in a deteriorated condition. On 27 February 1926 it was stricken from the list of naval aircraft.

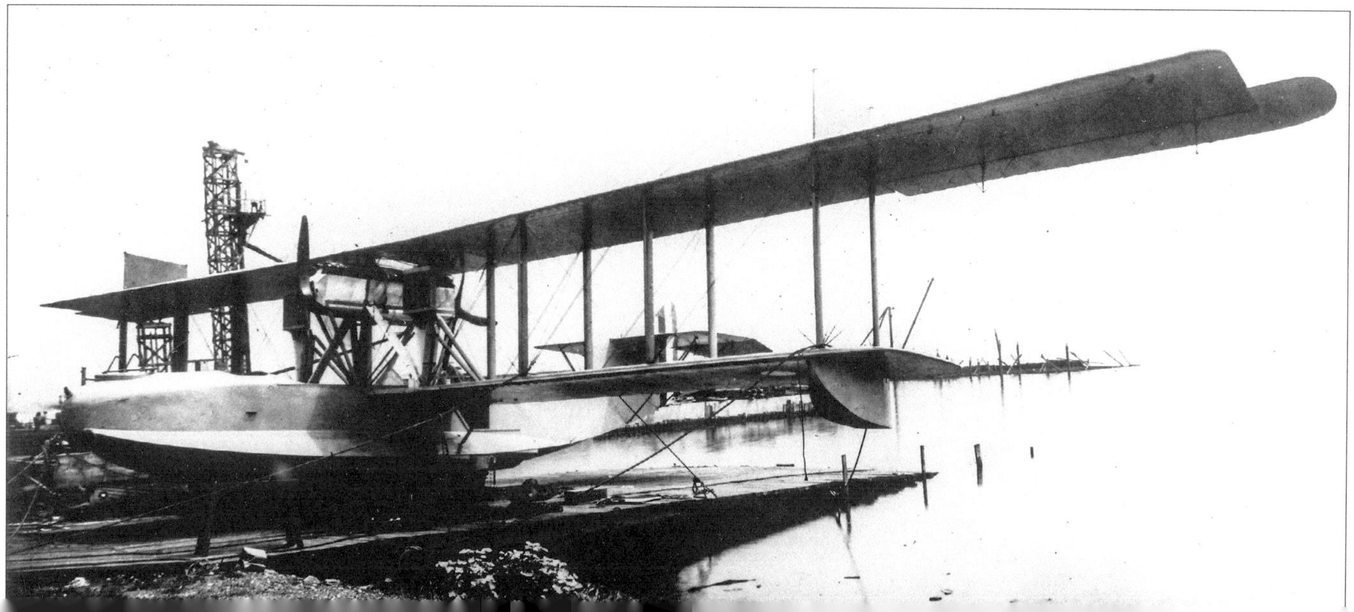

Above: A-3871 shows an anomaly in that the aircraft number "12" bears no relationship to the Bureau No. This may represent a replacement rudder from another aircraft, or a change in the way the aircraft numbers were allocated. This photograph was taken at St Louis Bay, Haiti, on 1 January 1924. The Squadron was on its way to Culebra, Puerto Rico. Note the gunnery pennant on the bow.

Below: F-5-L under construction.

Above: Davis recoilless gun fitted on an F-5-L.

Below: F-5-L A-3882 on display, most probably at the 1919 Aeronautical Exhibition, NY. The small floatplane is the Loening Kitten.

Above & Left: A-3882 on display at the Smithsonian. Note the Aeromarine floatplane in the background.

This Page: A-3882 in storage at Silver Hill, Maryland.

A-3882 was built at the NAF with half of the hull not planked and open for inspection. As at 30 September 1918, A-3579 to A-4035 were reported as 0-80% complete. Expected completion was 10.18 to 06.19. A-3882 was constructed with half the hull open for inspection. On 21 February 1919, A-3882 was ordered to be shipped to the Aeronautical Exhibition, NY. In July 1920 A-3882 that had "been on exhibition at various aero shows during the last two years and is complete in every detail and fully equipped with machine guns, bomb racks, etc.," was to be presented to the Smithsonian Institute. On 2 October 1920 the boat was shipped to the Washington Navy Yard and received for the Smithsonian Inst. On 8 September 1921, the flying boat was left at Hampton Roads. Engines 03219 and 04573. It was reported as unfit for service due to the length of time it had spent in storage on 4 December 1924. Repair and replacement would not be economical. To Smithsonian for display. Aircraft and engines to be stricken and be allowed remain for display purposes. 17th saw the boat officially stricken. Was displayed in Smithsonian but is presently in storage at Silver Hill, Maryland.

Above & Left: A-3882 in storage at Silver Hill, Maryland.

Above & Right: A-3882 in storage at Silver Hill, Maryland.

Above & Below: A-3882 in the Smithsonian's storage facility at Silver Hill, Maryland.

Above: F.3 N4009 was presented to Canada post-war as part of the Imperial Gift and was registered as G-CYDH on 11 July 1921, and was struck off charge on 11 January 1923. Aircraft with the registration prefix G-CY were Government (Air Board) owned aircraft

Above & Above Right: Canadian F.3 G-CYBT at rest.

Above: G-CYBT on 14 August 1921.

Above: Beached F.3. This machine bears cockades under the lower wing.

Above, Below, & Facing Page: Curtiss H-16 N4934 ended up as a house boat for these Naval Cadets. N4934 was delivered into store at AAP Norwich on the week ending 3 October 1918 and later scrapped with so many other boats in this batch. It survived for a number of years. Flying boat hulls were sometimes saved and used for a houseboat but too many were left to rot before their historical value was recognised. The Supermarine Southampton hull in the RAF Museum, Hendon, was recovered as a houseboat and saved.

N-4934

193

Above: Short air-yacht G-EAQT (ex-N4019). The flying boat first flew in this conversion on 28 May 1920. Shipped to Sydney, Australia, for businessman Lebbeus Hordern, it never flew again.

Above: The Netherlands interned aircraft of the waring powers that landed or were forced down in their territory. These included a number of seaplanes. Curtiss H-12 8689 is shown in Netherlands hands after recovery.

Below: N177 was built by Short Brothers with a metal hull utilising the aerostructure of the F.5. The bottom of the hull is painted with a white anti-fouling paint. The serial has a white outline on the natural metal hull.

Below: Canadian F.3 G-CYEN (ex-N4015). Registered on 21 August 1922 it was apparently little used, being reported as having crashed at Victoria Beach, Manitoba, on 8 September that year and was not repaired.

Above: Erecting a Spanish F.3.

Above: The cut-out at the centre of the wing can just be discerned on this photograph identifying this Portuguese Naval air service *Large America* as an F.3.

Above & Below: N177 was built by Short Brothers with a metal hull utilising the aerostructure of the F.5. The bottom of the hull is painted with a white anti-fouling paint. The serial has a white outline on the natural metal hull.

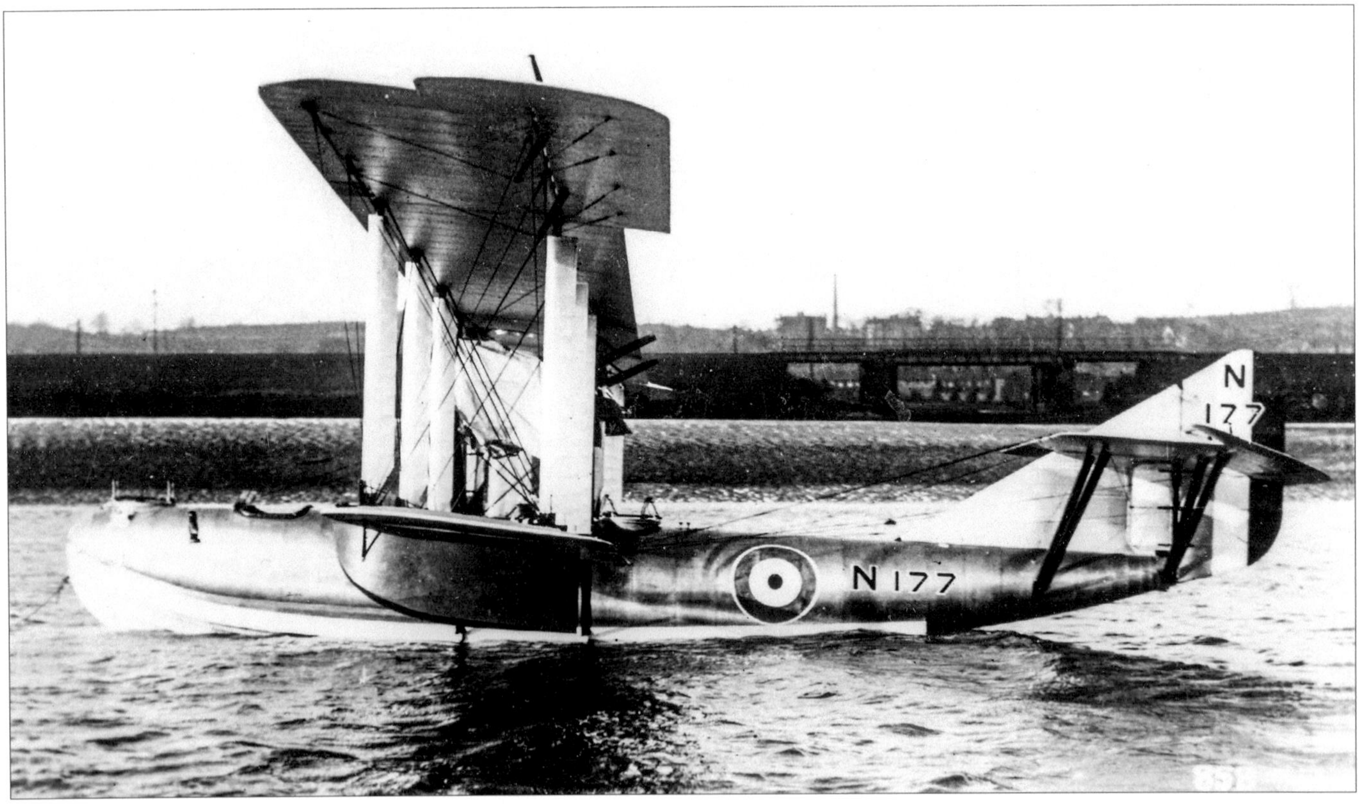

Above: N178 with experimental Saunders hollow hull. Saunders was given a contract for two experimental hulls to take F.2A wings. The Admiralty Technical Department report for the fortnight ending 12 June 1918, noted that one hull was finished and would now be fitted with standard F.2A wings and fittings. The following report noted that the progress on the second hull was satisfactory. What serials were allocated to these two experimental hulls were are unknown, but N178 may well be one of these.

Above: Possibly the first Japanese Navy Short-built F.5. This boat does not have any national markings applied as yet. The rear of the hull bears the Short Brothers logo and the c/n S.347.

Above: Running up the engines after assembly on a Japanese Navy F.5.

Above & Below Right: Japanese Navy F.5 serial 7. The hull appears to be black to the top decking. All other fabric surfaces appear to be V.84 aluminium.

Curtiss H-12 Convert 8689 was shot down by Germans and force to alight off the northern tip of the island of Vlieland on 4 June 1918. The machine managed to be run aground but before the crew could set the machine on fire they were prevented by the coast guard stationed on the island. The crew were all interned and the boat was impressed in the *MLD* as L.1. This day also saw Capt Dickey's F.2A N4533 shot up and set on fire on the water after alighting in Netherland's waters.

On 2 October 1918, F.2A N4551 of No.232 Squadron, RAF, was one of a flight of seven that flew along the Dutch coast within the Dutch territorial waters and were fired on by coastal batteries. Forced to alight, the boat was beached at Noordwijk. The crew were prevented from setting the boat on fire. This F.2A was taken into MLD service as L 2. According to a contemporary report three Canadian and two British men comprised the crew (Lt J.C. Stockman, 2/Lt T.N. Enright[56], 2/Lt W. Pendleton, 1AM H.L. Curtiss, and 3AM W.A. Mitchell). All were interned. Lt Stockman was injured in the incident, the rest were unhurt.

Above: Japanese constructed Hiero F.5.

Below: Argentinean F-5-L serial B-8. Note markings under lower wing.

Above: Argentinean F-5-L on board the *Dedalo*. Note the wing panels on the deck.

Portugal

In May 1920 two F.3 boats were reconditioned by Fairey Aviation Co Ltd at Hamble before being flown to Lisbon for the *Servico de Aviçã* (Naval Aeronautical Service). They were in service until 1922. One was used for the first flight from Lisbon to Funchal on Maderia Island in 1921. This 7 hour 40 minute flight was captained by Cdr Sacadura Cabral, with Capt Gago Coutinho as navigator. Cabral had flown the first of the boats from the UK to Lisbon. Coutinho was over 60 years of age and a well qualified navigator. This flight served to give them experience in navigation for the proposed first South Atlantic crossing the following year in a Fairy IIID.[57]

Spain

When the Riff War began in Spanish Morocco the Army and Naval air service were poorly equipped and there was a rush to acquire new equipment. The Army purchased Bristol Fighters, De Havilland 9 and 9A bombers from the ADC. Although

Above & Above Right: Brazilian F-5-L boats.

the war was mainly an Army affair, the Navy purchased 11 Felixstowe flying boats, variously referred to as F.3 and F.5 models from the ADC. Photographs show a cockpit canopy which would confirm that they were the F.3 model. They were erected, inspected, crated and shipped to Spain. As facilities were not available for their operation, the wings and tail sections were stored ashore while the hulls were placed on the davits of two old warships in Barcelona harbour, exposed to the sun and dew, a treatment for which they had not been designed. The first boat was wrecked while attached to its trolley in a gale the night before it was due to fly for the first time. For the story how they were brought into service see Norman Macmillan's *Freelance Pilot,* Heinemann, London, 1937.

Argentina

The Argentine Navy purchased eight F-5-L flying boats in 1921 that bore the serials B1 to B8. They were armed and entered service with the *Escuela de Aviación Naval* in 1922. With Dornier Wal flying boats they formed the first operational unit, the *Escaudrdilla de Patrulleros,* in 1922. In 1931 they were withdrawn for use as trainers, the last being written off in 1935.

Brazil

Brazil also operated the F-5-L, 14 being acquired in 1923 and given serials 1 to 5, and 21 to 23, and later 311 to 316 and 321 to 326. They served until 1930.

Peru

In 1920 the US Navy sent a mission to Peru that led to the established a seaplane base at Ancón with two Curtiss MF flying boats. After one of the MF boats crashed during a visit of the US Fleet to Peru in 1921, a twin engined flying boat was given to the Peruvian Navy. This boat is thought to have been a Curtiss H-16 and this aircraft had the shortest life of any in the Navy, crashing on 20 June 1921, after delivery the same day.[58] Some sources state that this was an F-5-L, however research by Dan Hagedon noted that a "Flying boat N-16 arrived on board the SS *Urubamba* 13 February 1921, ex-U.S. Government. It crashed on its first flight on 20 June 1921." It is thought that the N-16 is a misquote for an H-16. Peru's Navy was to operate two F-5-L boats from 1920 to 1924 along with De Havilland D.H.9 bombers and Avro 504K trainers.[59]

Post-War Civil

The Gosport Aviation Co on the Solent was building the F.5 and, according to H. Penrose, Porte had announced that he would be willing to join the company after the end of the war.[60] *Flight* for 15 December 1919, devoted an article – "Some Gosport Flying Boats for 1920" to the company's proposals. The article noted that the late John Porte had joined the company in August 1919 as Chief Designer, and had produced several designs based on his successful types in the late War modified to suit commercial requirements. The types are illustrated by drawings and their Felixstowe lineage can be clearly seen. The G.9 was a triplane flying boat based on the Fury and featuring the 600-hp Rolls Royce Condor engines that were to have originally powered the Fury. They were arranged with the two outboard engines as tractors and the central one as a pusher.

With his death on 22 October 1919, at only 35 years of age, Porte's ideas for civilian versions of the F-boats died with him. Not one Gosport design was ever built. The North Atlantic was conquered by a flying boat in 1919 before he died. This was the USN NC-4, a design that was radically different from those designed by Porte, but one which he would have been aware of given the constant communication between Felixstowe, Curtiss and the USN. This is not to say that the Porte boats had no influence on civil aviation.

Major W.T. Blake departed Crydon in a modified D.H.9 G-EBDE in an attempt to fly around the world on 25 May 1922. The original D.H.9 was lost in a crash and he continued his abortive world flight in a variety of aircraft. F.3 G-EBDQ was reportedly shipped to Canada for this part of his trip but was not used and later sold in Canada in 1922.[61]

Short Brothers Ltd had hoped to sell their F.3 and F.5 boats on the civilian market but had to compete with the low prices from the Aircraft Disposal Co (ADC). In their advertisements Shorts offered their F.3 as an adaption of this machine for passenger work that turned it into a luxurious yacht. A cabin

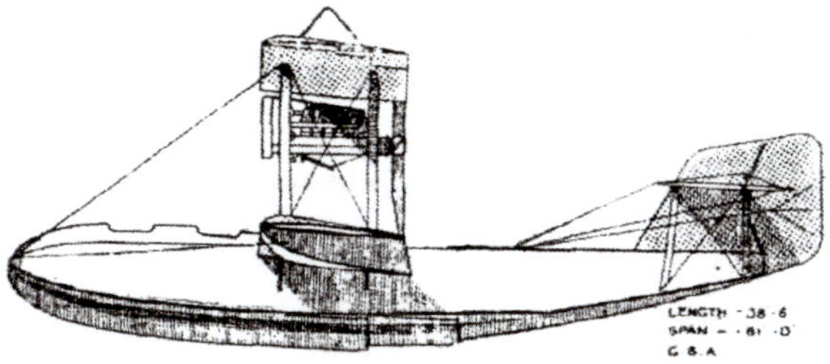

The Gosport G 8a type flying-boat, of which the G 8 is a slightly larger model

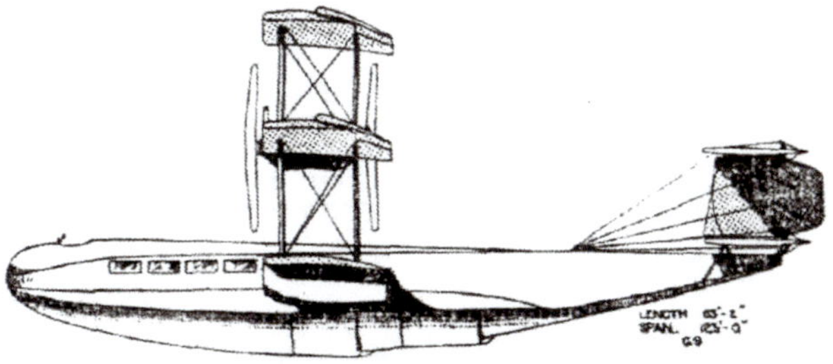

The Gosport G 9 triplane flying-boat. Fitted with three engines, it is intended for long-distance cargo work

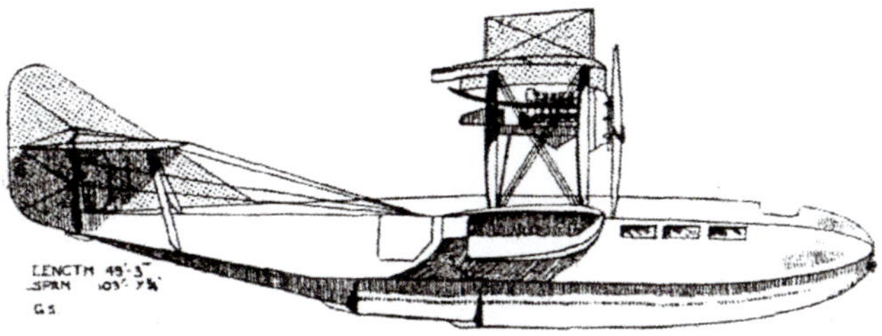

The Gosport G 5 flying-boat, for passenger or goods

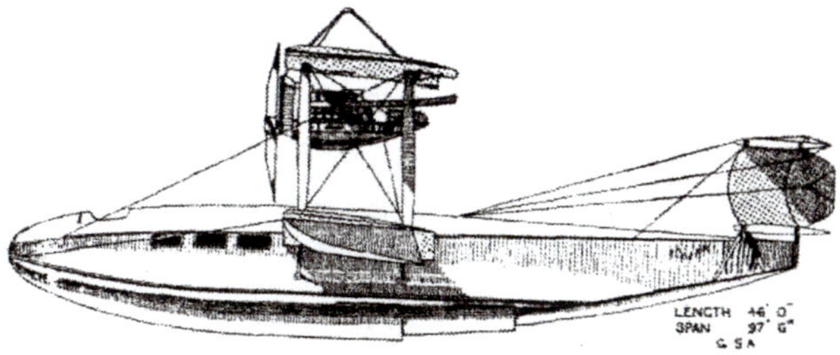

The Gosport G 5a, which is a slightly smaller modification of the G 5

Above: Drawings of some of the Gosport designs.

space 9 ft x 4 ft 6 in was fitted out as a lounge for four to six persons. If less comfort and luxury was required then a greater number of passengers could be carried. Two machines ended up on the British civil register as G-EAQT (ex-N4019) and G-EBDQ (ex-N4177), the latter, Blake's machine, referred to above.

G-EAQT was converted to an air-yacht by Shorts and flew in this form on 28 May 1920. Lebbeus Hordern imported G-EAQT to Australia together with spare parts. In Australian documents it is referred to as a Short F.5 flying boat. The hull was launched but the machine was never erected. In 1923 the Superintendent of Aircraft received a report on the machine. The wings were still in their packing case. The hull was thought to require a considerable amount of repair before the machine would be considered airworthy as all the joints had opened up and the timber had warped badly.[62] Hordern later donated the two Eagle VIII engines to the RAAF in 1927. According to legend the hull became a shelter for fishermen. The British registration had been cancelled in January 1921.

Some sources quote two F.3 boats used for an aerial service in Tasmania but no such service existed.

The Aircraft Disposal Co was still able to offer the following airworthy aircraft for sale in 1924:

Felixstowe F.3, Martinsyde F.4, D.H.9 and 9A, Bristol F.2B, Sopwith Snipe, Avro 504K and Parnall Panther.[63]

Aeromarine

In the Philadelphia Navy Yard were stored considerable numbers of F-5-L flying boats for which the Navy had no immediate need. 20 of these flying boats were offered for auction on 22 September 1919. Paul G. Zimmerman of the Aeromarine Plane and Motor Co was of the opinion that their sale at the low price would flood the market. He put forward a scheme whereby they would be given to cities and municipalities who would operate these after conversion to civilian passenger carriers by members of the Manufacturer's Aircraft Association. Hunsaker replied that while he had read the proposal with "considerable interest" he was of the opinion that it would take an act of Congress to allow for the disposal of the boats in the manner presented by Zimmerman. This scheme came to nothing.[64]

In March 1920, Zimmerman again wrote to Hunsaker. He stated that the company was just working on their first conversion of an F-5-L into a commercial machine. This was purchased for around $6,500. Five machines were contemplated for conversion. "We are satisfied that as soon as the first machine is flown much publicity will be given to it and from this publicity inquiries will result." Zimmerman wanted a full set, some 2,000 drawings for the F-5-L. In reply to this request, the Bureau of Supplies and Accounts noted that six F-5-L machines had recently been sold at a price of $10,850 each and it was understood, unofficially, that Aeromarine would modify the aircraft. The Bureau thought that Aeromarine should pay for the drawings.[65] What the result was is not known, however about eleven F-5-L boats were converted by Aeromarine Plane and Motor Co to a passenger configuration. Given the designation Aeromarine 75 they were capable of carrying ten passengers. The aircraft were

Above & Right: The *Sampaio Correia* over New York on leaving for the flight on 17 August 1922.

Above: The crew of the *Sampaio Correia*. The H-16 was decorated with the US and Brazilian flags. Walter Hinton (second from the left) and his crew in New York on the day they left on their flight to Brazil.

Walter Hinton learnt to fly in 1917 at Pensacola and was always interested in the big flying boats that started to arrive on the US entry into WWI. He had become interested in aviation when he had served on the USS *Seattle* that had a catapult and six Curtiss N-9 floatplanes. He piloted the NC-4 on its flight across the Atlantic.

He was in the news again in 1920 when the balloon he and two other officers from Rockaway were in was carried into Canada, coming down near Hudsons Bay. Given up for lost the crew reached civilization 31 days later. Hinton resigned his commission on 1 January 1922, to become the pilot of the Navy surplus Curtiss H-16 *Sampaio Correia* to undertake the first flight between New York and Rio de Janiero. The flight was to celebrate the friendship between the US and Brazil on that nation's centennial. The flight was sponsored by the *New York World* newspaper.

On 22 August they left Nassau for Port au Prince, Haiti. Arriving near Cuba in the dark on a moonless night they were well short of their destination and Hinton decided to set down near the US Naval Station at Guantanamo Bay, Cuba, which was nearby. Hinton knew the area well and finally located the Windward Point Light and lined up to alight. Unfortunately the light he chose was the USS *Denver* and the boat hit hard in the open sea rather than the protected waters of Guantanamo Bay, tearing the bottom off the boat. The *Denver's* crew had seen the lights of the seaplane but thought it was a shooting star. The ship's captain remembered the proposed flight of the *Sampaio Correia* and investigated to find the crew on the wreck of their aircraft. The five men survived the crash without injury but the boat was a total wreck.

The USN provided another surplus H-16 that Hinton christened *Sampaio Correia II*, and the flight recommenced from Pensacola on 3 September. After many adventures the machine arrived at Rio de Janeiro on 9 February 1923. It had taken nearly six months to make the 8,500 mile trip.

As a member of the Hamilton Rice expedition of 1924–25 Hinton flew a Curtiss MF flying boat deep into the Brazilian jungle.

He was the last surviving member of the NC trans-Atlantic flight. He died on 28 October 1981.

Above: The sad end of the first *Sampaio Correia*.

operated by the company's own airlines: Aeromarine Airways and Aeromarine West Indies Airways, flying from Key West to Havana; New York to Atlantic City, and Cleveland to Detroit. They are credited with carrying the first international air mail from the USA.

Aeromarine literature stated that this

Aeromarine aerial cruiser, which has ample accommodation for ten passengers, exclusive of pilot and pilot's mechanician, has been designed for aerial voyages from New York to Asbury Park Atlantic City, Norfolk, Washington, Baltimore, Southampton, New Haven, New London, Newport, Miami and other points on the Atlantic.

The trip would be made "in appointments that even the custom-built creations of the Automobile world, this new Aeromarine cruiser has a comfort all its own."[66]

The "Aircraft Yearbook" for 1921 listed two F-5-L Cruisers with Aeromarine Sightseeing & Navigation Company (merged with Aeromarine West Indies Airways Inc), and six F-5-L Cruisers with Aeromarine West Indies Airways Inc. In the 1924 edition it lists five F-5-L thirteen-passenger craft with Aeromarine Airways Corp.

In February 1921 the Bureau of C&R recommended the sale of two F-5 flying boats to the Aeromarine Plane and Motor Co. They were to be delivered as is and not removed from their crates for overhaul. If any surfaces required covering, this was to be furnished separately.[67]

Aeromarine Airways lasted until 1924 when it ceased operations. The full story of his pioneer airline may be found in "Aeromarine Airways – Its Aircraft and History," by D Koch, in *Skyways, the Journal of the Airplane 1920-1940,* No.52, Oct 1999.

Specifications Aeromarine 75

Source	1	2
Span upper	103 ft 9 in	103 ft 9¼ in
Span lower	74 ft 4 in	74 ft 4 in
Length	49 ft 4 in	49 ft 3¼ in
Height	18 ft 9 in	18 ft 9¼ in
Chord	8 ft	8 ft
Gap	8 ft 10.5 in	8 ft 10½ in
Incidence	4°	3° 40'
Dihedral	1° 30'	1° 30'
Aerofoil	RAF 6 Mod	
Areas in Sq. Ft.		
Wings	1,278.0	1,394
Ailerons (2)	119.0	118
Wing top (inc ailerons)		848
Wing bottom		546
Elevators	55.3	
Horizontal Stabilizer	120.9	
Rudder	33.5	
Fin	34.7	
Non-skid Fins	31.1	30
Total Weight		13,659.5 Lbs
EmptyWeight		8,456 Lbs
Fuel Capacity		230 US gallons
Speed		50/85 mph
Climb to ft		2,600/10 min
Endurance		Approx 5 hours

1. Model 75 General Characteristics, Aeromarine document. Copy in NASM Smithsonian Institute File AA-045270-01 Aeromarine 75 (F-5L).
2. D Koch, "Aeromarine Airways – Its Aircraft and History" *Skyways, the Journal of the Airplane 1920–1940*, No.52, Oct 1999, P.9.

Chapter 6 Endnotes

1. Minutes of 17th Air Department Weekly Progress Meeting. 04.09.1917. TNA AIR1/270/15/226/113.
2. Minutes of 18th Air Department Weekly Progress Meeting. 11.09.1917. TNA AIR1/270/15/226/113.
3. Minute 143398: Director of Air Services 22.12.1917. TNA AIR1/664/17/122/695.
4. Sheet headed: SEAPLANE TRAINING with above. This policy was agreed to continue at a Meeting on 14.01.1918. It was decided that Calshot required six (6) *Large Americas* for the Large America School and three (3), together with 12 float seaplanes, at the Gunnery School. Minute: 143635. *Ibid.*
5. Minute: C Courtney(?), N.1. 03.01.1918. He was asking for confirmation of the policy. A school floatplane was considered to be required. *Ibid.*
6. According to William G Carr the "pigeoners" were busy men as they were used as Jack-of-all-trades. They looked after the lifebelts, leather jackets and trousers of the flight crews. The Felixstowe War Flight was always very short of kit and they had to be passed on from man to man. They kept the engineer's tool kits, looked after emergency rations, sea anchors, the cameras and in their spare time acted as messengers. This was in addition to looking after and training the pigeons. The Germans usually released the birds that fell into their hands. When a young bird proved reluctant to fly in daytime, perching on some part of

Above: View of one of the cabin modifications used by Aeromarine for the F-5-L. This appears to be an early modification as most had a more rounded bow.

Below: *Santa Maria* has the route and time proudly displayed on the hull.

Above: Another view of *Santa Maria* with the route and travel time proudly displayed on the hull.

Above: Close-up of *Santa Maria*. The boats were kept in good condition and must have been an impressive sight in the air.

Above: Front view of the *Santa Maria*.

Above, Below, & Facing Page, Top: *Santa Maria* in another colour scheme. The raised pilot's cockpit under the wing and the doors at what was the side gun positions are readily seen in these photographs.

Above: *Santa Maria* powering to get up onto the step.

Above: *Buckeye* in flight. The colour scheme is black and white.

Above: *Columbus* in flight over Bimini in the Bahamas, late 1921.

Above: *Buckeye* low over a smog and fog covered Cleveland in the summer of 1922.

Above: *Nina* carries a cine-cameraman in the front cockpit.

Above: A-3600 the "Denby Ship" was converted to a passenger configuration at the NAF. Note the containers under the lower wings. In November 1918 this machine was reported as 60% complete. By 8 August 1921, this boat was converted to a passenger ship for the Secretary of the Navy and had been flown at the NAF with ten persons. On 20 September 1921, was shipped to Anacostia. On 30 December 1921, Lt F.H. Becker reported that controllability was bad and requested an investigation. In June 1922 Anacostia recommended it be replaced by a standard F-5-L as "its carrying capacity and controls are below that of standard F-5-L." It appears that it was placed in storage at the NAF and offered for sale in 1924. It was stricken on 2 July 1925. (G648)

Above: This modification did away with the gunner's position in the nose replacing it with a hatch that opened for mooring.

Above & Facing Page, Bottom: The cylinder under the wing of this converted F-5-L appears to be a storage container and while its purpose is unknown it appears to be the same as A-3600. (NAS Anacostia photographs 1922).

Above: The *Santa Maria* was, according to the original caption, "the largest commercial flying boat in America arrives here (New York) from a record flight from Havana. The *Santa Maria* holds the record for continuous flights." Cmdr J. McAtes is releasing carrier pigeons with messages for Secretary Denby on their arrival.

Above: Interior of an Aeromarine 75 showing the passenger accommodations.

Facing Page, Bottom: Modifying an F-5-L at the Aeromarine works 06.11.1920.

the machine, punishment soon followed. "Birds that refused to do their duty had their commission cancelled and were killed and eaten in the form of very tasty pigeon-pie." Carr, W.G. *Out of the Mist.* P.32.

7. Hallam, T.D. *The Spider Web.* P.91.
8. Robinson, A. "The Recognition of First War Flying Boats", *Cross & Cockade Great Britain,* Journal, Vol.11, No.1, 1980, P.21.
9. Report: pencilled note on this report states it is from Hope-Vere and dated 14.03.1917. TNA AIR1/305/15/226/153 Part II. According to the 1925 *A Manual for Rigging Aircraft,* (Air Publication 1107, HMSO, London), under the heading "Floats and Hulls," *Seaplane floats and hulls of flying boast require special consideration and treatment on account of their very exacting conditions of service. Only timber of the best quality should be used, and all fasteners must be of yellow metal, such as copper nails or brass screws, rod or wire. White boat nails cannot be turned or clenched, screws should be used in their place. Dead nails should not be used.*
10. *Ibid.* It is assumed that the reference to Felixstowe hulls means he wanted to modify the boats to H-12 Converts.
11. Quoted in Penrose, H. *British Aviation – The Great War and Armistice.* P.202.
12. Force Cmdr, US Naval Forces Operating in European Waters, to Naval Cmdr. Subject: "Loss Flying Boat N4510 – Report On." NARA RG 45 Box 154.
13. According to J.M. Bruce the layers were: inner 1/8in cedar and outer 3/16 mahogany, 5 in wide planks.
14. Oswald Short wrote that the F.2A was "almost unmanageable in turbulent air." Although Shorts was a main builder of the F.3 and F.5, according to Oswald, Porte was so biased towards flying boats "that he would not have a Short seaplane at his station." He stated that Felixstowe was the main base for official trials and his seaplanes were not given a fair assessment of their good qualities. A considerable amount of testing was actually done on the Isle of Grain. Oswald was also opposed to the Admiralty Air Department. Here his criticism of Government designs carries more weight as none of the A.D designs were really successful. Quoted by Penrose, H. *Op. Cit.* P.237.
15. *Sixty-First Report of the Controller, Technical Department, for the fortnight ending 21st August, 1918.* Copy in RAF Museum J.M. Bruce Collection, Box 25.
16. Grain Test Report No.NF240. "Wright Stabiliser fitter to H16 Boat No. N.4892" TNA, Isle of Grain (Test Depot) Reports, N.F. Series Nos. 109-219 consecutive. 28.02.1918 – 04.01.1925.
17. Superintendent Constructor of Aircraft, Curtiss Aeroplane Co, Buffalo to Bu C&R . 26.07.1917. NARA RG72 Box 385 File O-ZF-2 Vol.1.
18. Minutes, Air Department Progress Committee, 12.06.1917. TNA AIR1/270/15/226/113.
19. Bruce, J.M. "Porte and his Felixstowe Flying Boats," *Cross & Cockade 25th Anniversary Seminar Papers*, Cross & Cockade International, UK. P38.
20. Carr, W.G., identifies the submarine as the UC-1.
21. Hallam, T.D. *Op. Cit.* P.155.
22. Minutes, Air Department Progress Committee, 12.06.1917. TNA AIR1/270/15/226/113.
23. "F-Boats – a postscript by an ex-pilot," *Flight,* 20.01.1956, P.86.
24. Report N.F.135. Test Depot, Isle of Grain. "Structural Defects." 22.03.1918. TNA, Isle of Grain (Test Depot) Reports, N.F. Series Nos. 109-219 consecutive. 28.02.1918 – 04.01.1925.
25. Brig Gen Groves, Director Air Division, Admiralty, 15.10.1918. TNA Mun 8/28.
26. Brig Gen O. Swann, DAC Orkneys & Shetlands, to Vice Adm Northern Patrol, 23.06.1918. TNA AIR1/643/17/122/269.
27. The same as the RN carried out in the Straights of Dover where the Allied naval forces laid down a mine barrage across the Straights patrolled by trawlers and drifters. The Otranto barrage was even less successful than the one across the Straights of Dover. U-Boats continued to use the Straights, particularly at night.
28. Copy Report in AIR1/305/15/226/153.
29. Minutes of 1st Air Department Progress Committee Meeting of 15.05.1917. TNA AIR1/270/15/226/113.
30. Telegram: SNO Malta to Admiralty, 21.11.1917. TNA AIR1/305/15/226/153.
31. *Sixty-First Report of the Controller, Technical Department, for the fortnight ending 21st August, 1918.* Copy in RAF Museum J.M. Bruce Collection, Box 25.
32. RAF Museum J.M. Bruce Collection, Box 25.
33. Isle of Grain Report N.M.241 by Maj A.Q. Cooper.02.12.1918. TNA AIR1/1204/204/5/2616.
34. Minutes, Air Department Progress Committee, 17.07.1917. TNA AIR1/270/15/226/113.

35. Air Commodore C.R. Samson, "The Operation of Flying Boats in the Mediterranean." *RAeS Journal* Proceedings 60th session, 05.02.1925. P.121.
36. Rennie, J.D. "Some Notes on Flying Boats," P.123.
37. 61st Report of the Controller, Technical Dept for fortnight ending 21 August 1918. RAF Museum J.M. Bruce Collection, Box 25.
38. 70th Report of the Controller, Technical Department – Aircraft Production, for the three weeks ending 1 January 1919. P.5. Copy in RAFM J.M/Bruce Collection Box 25.
39. NAF to Bu C&R. "First F-5-L F/Boat – Completion of." 15.07.1918. NARA RG72 File O-ZF-9 Vol.1.
40. R.E. Bassler to J. Hunsaker RE: Visit to Philadelphia Naval Aircraft Factory. No date.
41. Letter B. Robertson to J.M. Bruce, 09.11.1955. RAFM J.M. Bruce Collection Box 42.
42. "U.S. Naval Airplane The Curtiss F-5-L Flying Boat," personal reminiscence of a flight mechanic from Pensacola from 1926. Unfortunately the manuscript is unsigned. Copy in NASM File AC-901400-01 Curtiss F-FL.
43. Minutes of 19th Air Department Weekly Progress Committee Meeting of 18.09.1917. TNA AIR1/270/15/226/113.
44. Grain (Armanent) Report No.33. "The Le Vaillant Speaking Tubes for use in Large Flying Boats." 01 Sept 1918.TNA A1/1201/204/5/2611.
45. "A Seaplane Endurance Test," *Flight,* 29.05.1919. P.692.
46. Report N.F.268a & N.F.268b. MAEE to Air Ministry. 21.02.21 & 11.03.21. TNA, Isle of Grain (Test Depot) Reports, N.F. Series Nos. 109-219 consecutive. 28.02.1918 – 04.01.1925.
47. Report MN271 in TNA AIR1/1204/204/4/2626.
48. "Flying-Boats for Scandinavia," *Flight,* 04.09.1919. P.1202. No mention of the second boat, N4041, and what prevented it from undertaking the flight is mentioned in the *Flight* report.
49. Canada received 64 Avro 504K trainers; 13 D.H.4 and 12 D.H.9A bombers; 13 S.E.5a fighters, two Sopwith Snipes and one each of a Bristol F2B Fighter, Fairey IIIC, F.E.2d, Vickers F.B.9 and Curtiss H-16.
50. It appears that N4178 and N4179 were not on the official acceptance sheet of the Canadian authorities which explains the different numbers given for the Imperial Gift machines.
51. Sturtivant, R. and Page, G. (*Royal Navy Aircraft Serials and Units 1911–1919*), *Op. Cit.* The Imperial War Museum Film Catalogue Item IWM 861"Fire Fighting with Aeroplanes," shows flying boats fighting a forest fire in Canada and includes shots of F.3 G-CYBT. (http://www.iwm.org.uk/collections/item/object/1060023292)
Catalogue Item 859 "In the Wake of Captain Cook" shows F.3 G-CYDI on a visit to Nootka Sound on Vancouver Island, circa 1922. http://www.iwm.org.uk/collections/item/object/1060023290
52. *The Aeroplane* Journal, 06 Nov 1942. This article refers to the paper that was read by Col the Master of Sempill, AFC, RAF (now Lord Sempill), before the RAeS on 3 April 1924. The article noted that the F.5 type had been superseded by the "newer P.5 basic design, then in construction by several firms in Great Britain." In fact the development of the P.5 remained only experimental and was produced in small numbers as the English Electric Kingston. See RAFM J.M. Bruce Collection Box 1. See also Owers, C.A. "The Phoenix P.5 Cork," *Cross & Cockade International,* Vol.45 No.1, 2014.
53. There is much confusion as to exactly how many F.5 boats the Japanese received from Britain. F/Off R. Vaughn-Fowler who accompanied the Mission to the Imperial Japanese Navy recorded his impressions and filed a report on his return to Britain. He listed all the aircraft supplied and under *F.5 Flying Boat 2. 375 h.p. Eagles*, lists 15 machines. Where he was not sure as to the exact number Vaughn-Fowler added a "?" to the number. There is no "?" with the F.5 listing. TNA AIR5/358.
54. "New F-5 Flying Boat." 17.03.1920. Copy in NASM File AC 901400-01 Curtiss F-5L.
55. Australian National Archives CRS A705/1 Item 114/6/28. With Quarterly Letter to 31.03.1929.
56. Henshaw, T. *The Sky Their Battlefield II*, Fetubi Books, UK, 2014. P.265, states that Enright was missing in action.
57. Barnes, C.H. *Shorts Aircraft since 1900,* Puntam, UK, 1967, states that Shorts supplied a Short-built F.3 to the Portuguese Government in April 1920. This machine was registered C-PAON.
58. E-mail from Dan Hagedon, 15 Nov 2015. No H-16 is recorded for Peru in Green, W. & Fricker, J. *The Air Forces of the World,* NY,

1958, which leads to the conclusion that the flying boat that was lost on receipt was an H-16.
59. "Early Peruvian Air Arms," *Insignia,* UK, No.17, 2002. P.4. A similar Report (dated 5 April 1924) to that stated that "Currently on hand at Ancon are two seaplanes similar to the F-5-L (engines are on hand but are not mounted)." E-mail from Dan Hagedon, 15 Nov 2015.
60. Penrose, H. *Ibid.* P.399.
61. Another source states that the F.3 never went to Canada and a D.H.9 G-EBDF was sent to Canada instead. *The British Civil Aircraft Registers 1919-1928,* Air Britain, UK, 1969.
62. Australian National Archives, CRS A705 Item 35/6/899.
63. Orde-Hume, A. *The Great War-Plane Sell off,* GMS, UK, 2005.
64. Hunsaker to Zimmerman, 26.11.1919. Hunsaker Papers, NASM Smithsonian Institute. Accession No.XXXX-0001
65. Zimmerman to Hunsaker, 31.02.1920. Also Memorandum, Hunsaker to Bureau S&A, 7.04.20, and Bureau's reply of 09.04.20. *Ibid.*
66. "The Aeromarine Navy-Cruiser". Aeromarine advertisement pamphlet. NASM Smithsonian Institute File AA-045270-01 Aeromarine 75 (F-5L).
67. Letter: Bureau C&R to Bureau S&A. Subject: *Sale of F-5's to the Aeromarine Plane & Motor Co.(Supply Officer's letter #NAF Record 4864 of 27 January 1921).* 10.02.1921. NARA RG72. Entry 53. Bureau of Aeronautics General Correspondence initiated in the Bureau C&R 1917–1925. Box 361. File RF-Z1.

Above: Argentinean F-5-L flying boats on board the *Dedalo*.

Appendix 4.1

Report N.M.200 Comparison Between F.2A Production Machines

N4510 – Contractor: May, Harden & May.
N4283 – Contractor: Saunders.
N4060 – H-16. Contractor: Curtiss Co, America.

		F.2A flying boat N4510. 12/02/18. 2 x Eagle 8 Rolls-Royce			F.2A Flying Boat N4283. 06.03.18 2 x Eagle 8 Rolls-Royce			H-16 (F.2A) N4060. 31.05.18. 2 x Eagle 8 Rolls-Royce		
Climb	**Height, Ft**	**Time, M-S**	**RofC Ft/Min**	**ASI Kts**	**Time, M-S**	**RofC Ft/Min**	**ASI Kts**	**Time, M-S**	**RofC Ft/Min**	**ASI Kts**
	0	0-0	600 initial		0-0	565 initial		0-0	586 initial	
	2,000	3-45	491	59	3-50	470	57	3-40	512	61
	5,000	11-00	330	56	11-45	326	57	10-20	390	61
	10,000	39-00	90	56	39-30	86	54	28-00	198	59
	Absolute Ceiling 11,900 ft Service Ceiling 9,000 ft				Absolute Ceiling 11,600 ft Service Ceiling 9,650 ft			Absolute Ceiling 15,000 ft Service Ceiling 12,500 ft		
Speed		**Speed in Knots**			**Speed in Knots**			**Speed in Knots**		
	2,000 ft	84			83			85½		
	5,000 ft	81			80			84½		
	10,000 ft	74			70			80		
Weights in Pounds	Machine light: 7,722 276 gal petrol, 11 gal oil, 6 hrs fuel full out 2,057 Crew four 720 W/T, guns, etc No bombs 585 Total weight 11,084				Machine light: 7,549 276 gal petrol, 12 gal oil, 6 hrs fuel full out 2,124 Crew four 720 W/T, guns, etc No bombs 585 Total weight 10,978			Machine light: 7,363 276 gal petrol, 11 gal oil, 6 hrs fuel full out 2,124 Crew four 720 W/T, guns, etc No bombs 472 Total weight 10,670		

Appendix 4.2 – Comparison of F.2A & F.3

	F.2A	**F.3**
Endurance at full power	5½ hrs with bombs. 6½ hrs without bombs.	6 hrs with bombs. 8¾ hrs without bombs.
Endurance at cruising speed of 60 kts	8 hrs with bombs. 10 hours without bombs.	8 hrs with bombs. 11 hours without bombs.
Speed at full power	85 knots	75 knots
Full War Load	Crew (4) 720 Bombs 460 4 Guns 195 W/T 135 Total 1,510 lbs	Crew (4) 720 Bombs 920 4 Guns 195 W/T 135 Total 1,970 lbs
Ceiling with full load	15,000 ft Weight total 10,500 lbs	10,000 ft
Climb at sea level	650 ft/min	350 ft/min
Expected life	6 to 8 months	6 to 8 months

Source: Report in AIR1/305/15/226/153.

Appendix 4.3 – Comparison of the Prototype F.5 to Production Machines

Aircraft & Condition	Serial	Quoted Power	Empty Wt in Lbs	Load in Lbs	Gross Wt in Lbs	Speed in Knots	Service Ceiling in Feet	No. of Report	Date
F.5 with ordinary load	Prototype N90 Streamline wires – 4 x 230-lbs bombs	2 x 345-hp Eagle VIII	8,023	4,245	12,268	89	11,400	N.M.165B	09/05/18
F.5	N4039. Streamline wires – No bombs	2 x 352-hp Eagle VIII	9,134	1,526	10,660	82.5	12,500	N.M.243	10/19
F.5 with light load	N4040 Cables – No bombs	2 x 352-hp Eagle VIII	9,100	3,582	12,682	76.5	6,800	N.M.248	05/19
F.5 with light load	N4040 Cables – No bombs	2 x 352-hp Eagle VIII	9,100	1,414	10,514	78	10,000	N.M.248	05/19
F.5	N4839* Streamline wires – No bombs	2 x 450-hp Napier Lion	9,620	4,250	13,950**	78	6,650	N.M.271	16/08/22

Source: J.M. Bruce to Sir A. Robinson, "Comments on Professor Robinson's Notes, Feb–June 1959." Copy in RAFM J.M. Bruce Collection Box 22. 3067. With additions from Isle of Grain Report N.M.243 and N.M.271, copy in TNA AIR1/1204/204/5/2616 and AIR1/1204/204/4/2626 respectively.

* It was necessary to hand over N4839 to the Development Flight before the trails were completed. Principle performance characteristics were obtained.
** This included 80 lbs of absorbed water.

The dates of the Trial Reports M.N.165-165C confirm them as belonging to the prototype. Further confirmation is given by the different hp of the engines.

All speeds listed in the Record of Peformances Chart were adjusted to a standard atmosphere. Thus the figures printed were seldom the same as those in the actual trial reports but were after adjustment directly comparable, at least in theory.

Calshot – Aircraft Strength at the Armistice			
Type	240 Squadron Nos. 410/411 Flights	210 T.D.S. B Flight	Calshot Group Workshop & Repair Depot
Floatplanes			
Campania	N2376, N2395	N2381, N2386, N2387, N2389, N2393	N2379, N2399
Short 184 (Maori)	N2985, N9108	N1623, N2951, N2952	N2975
Short 320 (Cossack)	N1707, N1709	N1403	N1366, N1398, N1402
Wright S.P. (Maori)	—	N1280, N1282, N1283	N2185
Wright S.P. (Renault)	—	—	N1281
Flying Boats			
Felixstowe F.2A	N4081, N4285, N4286, N4535	N4292, N4534, N4547	N4309, N4536, N4544
Curtiss H-12	N4334, N4337, N4349		
Policy changed for Calshot after the war and by early 1919 only 3 F.1A boats and 2 Short 184 floatplanes were on Group operations Strength and the Training School has become a Floatplane Training School. Apart from a single F.2A (N4309) all other aircraft were Maori powered Short 184 floatplanes: N1621, N1623, N2290, N2999, N9039, N9042, N9050, N9091, N9110, N9111, N9174, N9175, N9176 and N918. Typed chart in RAFM JM Bruce Collection Box 3.			

Above: The US went back to a boat type hull for their NC flying boats. The USN and Curtiss developed the NC boats as a way of delivering the machines by air, freeing up cargo space for other vital war products. What, if any, influence Porte had on the design is unknown; however, it is known that Porte was in contact with the USN Bureau of Construction and Repair until he terminated his service with the RAF.

As the line of Brandenburgs re-formed for the attack, death dived among them in the shape of a hurtling D.H.4

Left: The war over the North Sea did not get the same coverage as that on the Western Front in the pulp magazines between the wars. Here is a sketch from the story "Raiders of the North Sea" from *Air Stories*, Vol.4 No.4, April 1937, depicting combat between a dazzle painted Felixstowe boat and Brandenburg monoplanes. The W.29 floatplanes bear the intermediate straight cross national insignia as seen on many and are lining up to attack from astern. The Felixstowe boats would dive for the surface of the sea in order to prevent the enemy from attacking from below where they had poor defence.

Air Stories was first published by George Newnes in May 1935. The editor was T. Stanhope Springs. It should not be confused with the American pulp of the same name. It advertised itself as the "All British Magazine" on the early issues. *Air Stories* told of the air war from the British point of view. The stories have Editor's notes where RFC/RAF slang and terms are explained. The magazine also printed war-time squadron songs. As photographs were not included sketches were a prominent addition to the stories with usually a leader at the start of the story and additional drawings if the story was of length.

Air Stories promoted the fact that many of its authors were ex-RFC and RAF members. Men such as Capt W.E. Johns, Arch Whitehouse, Major Oliver Stewart and Milford Hyde. Their rank and squadron affiliations were often given in the by-line of their stories.

Above & Below: Sketches from F.2A Rigging Notes

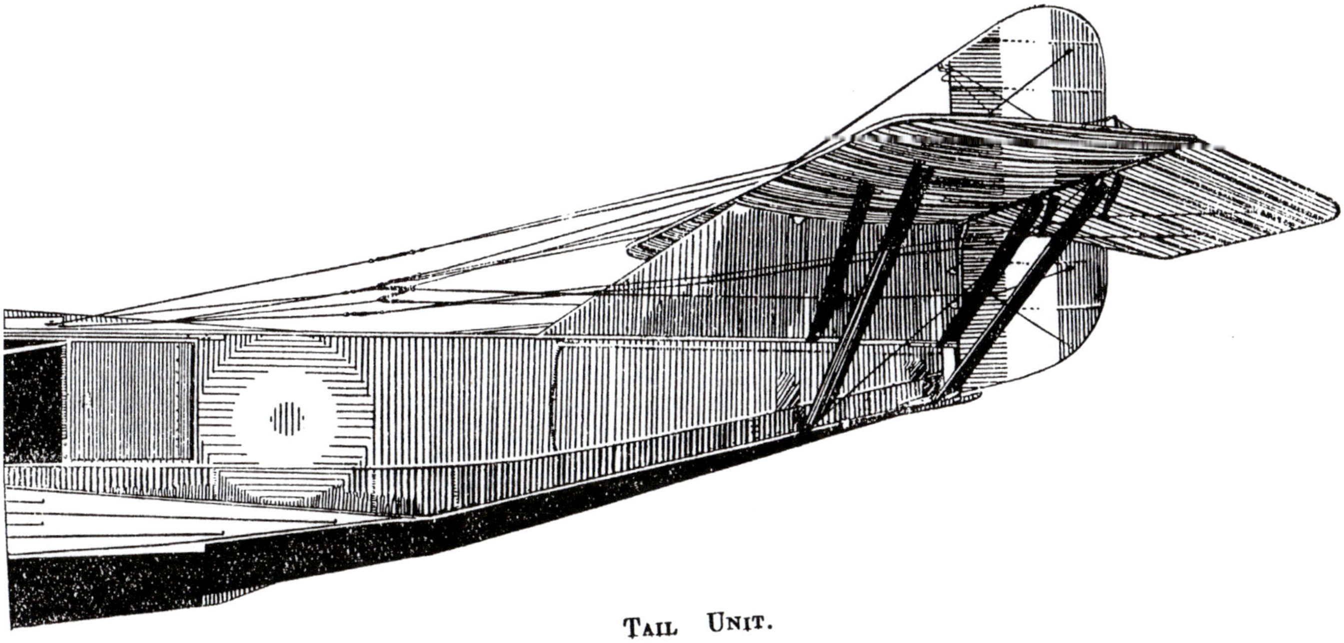

Tail Unit.

Above & Below: Sketches from F.3 Rigging Notes

Tail Unit.

Above & Below: Sketches from F.2A Rigging Notes

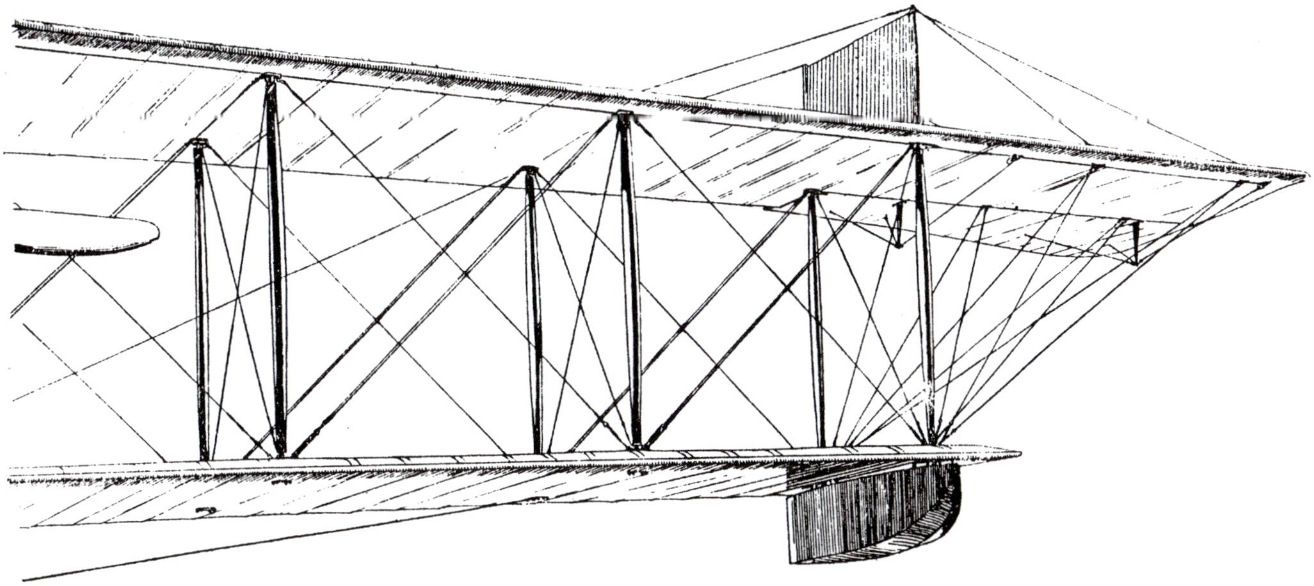

PORT PLANES, SHOWING WIRING.

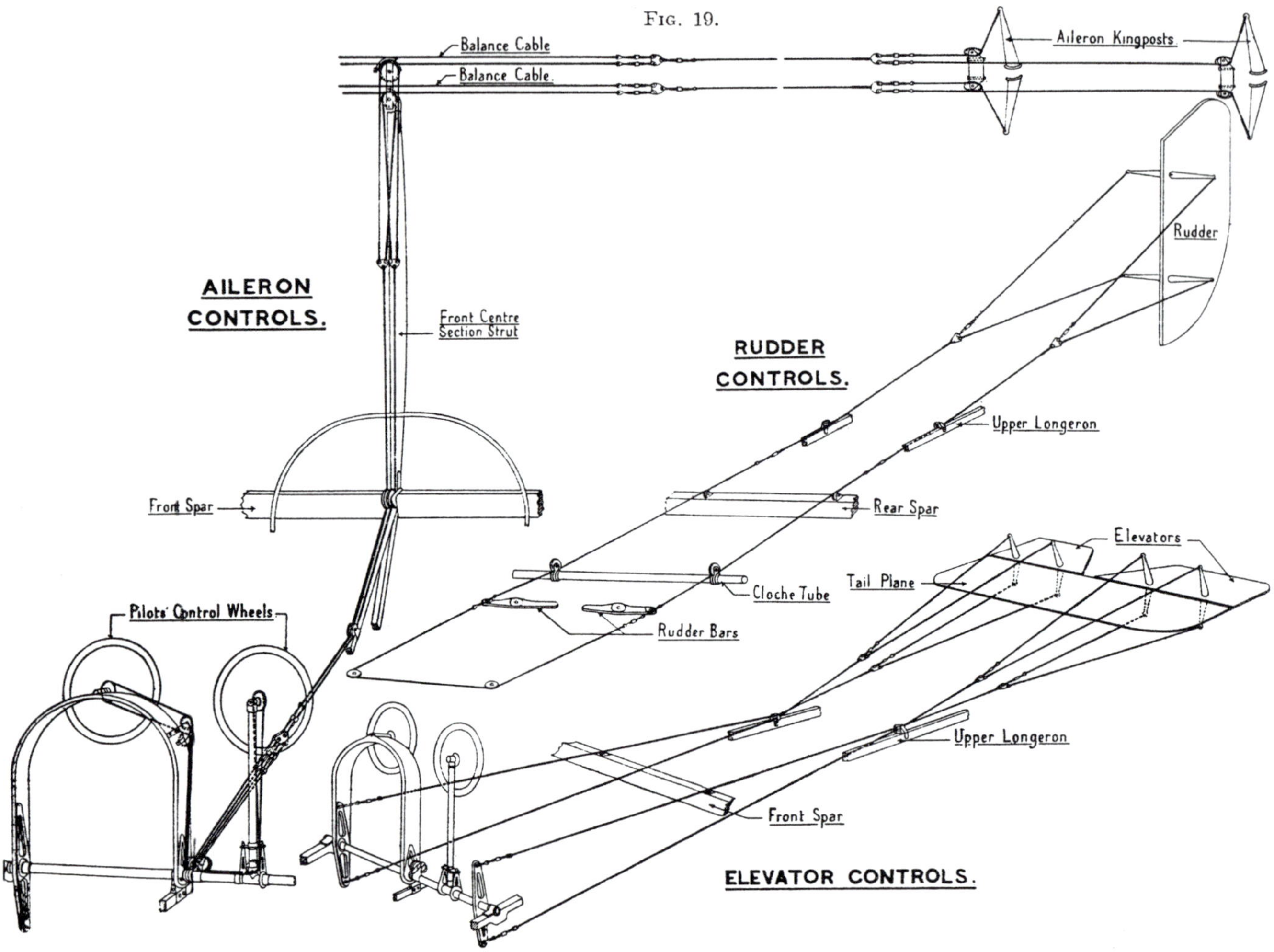

Above: Sketch from F.2A & F.3 Rigging Notes

Felixstowe F.3 in Spanish service.

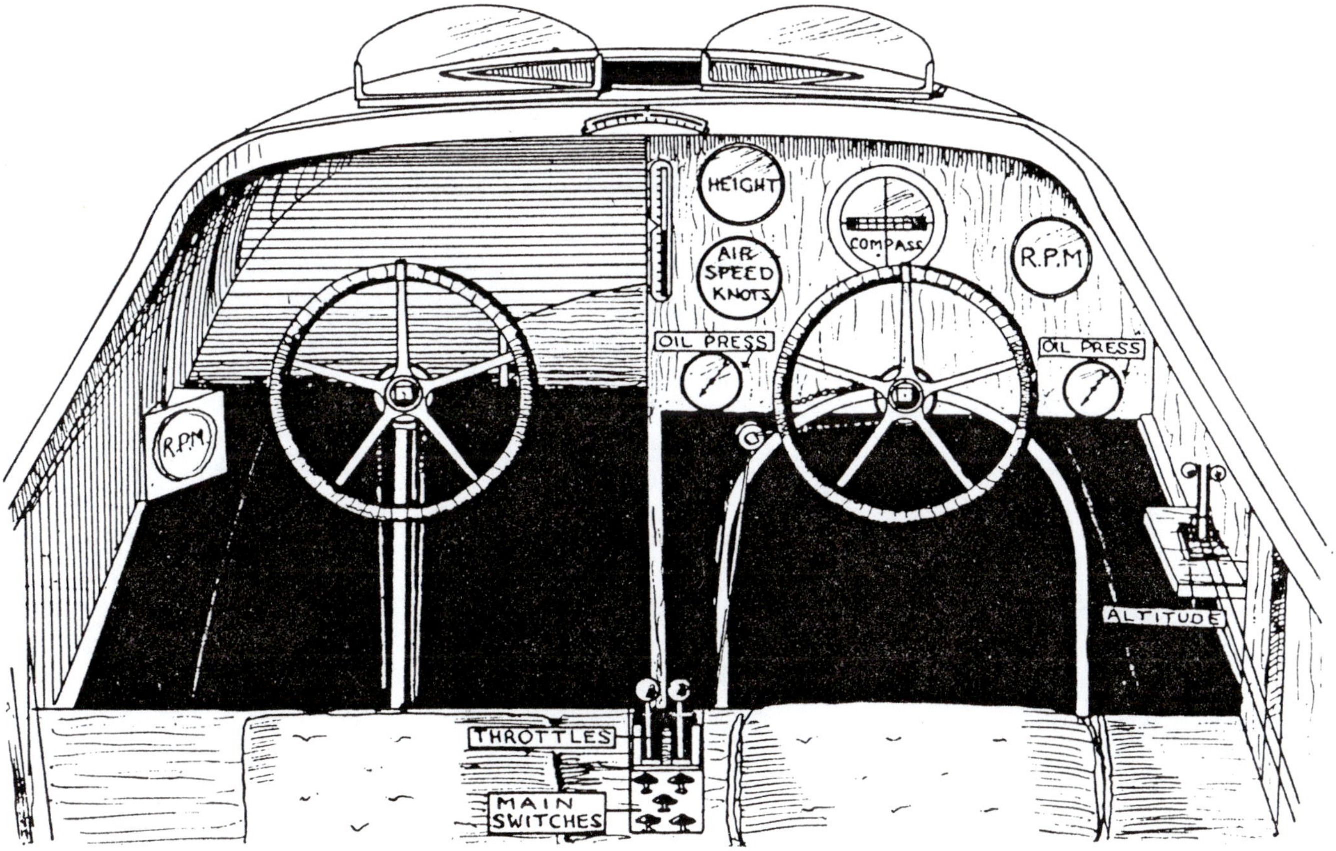

Above & Below: Sketches from F.5 Rigging Notes

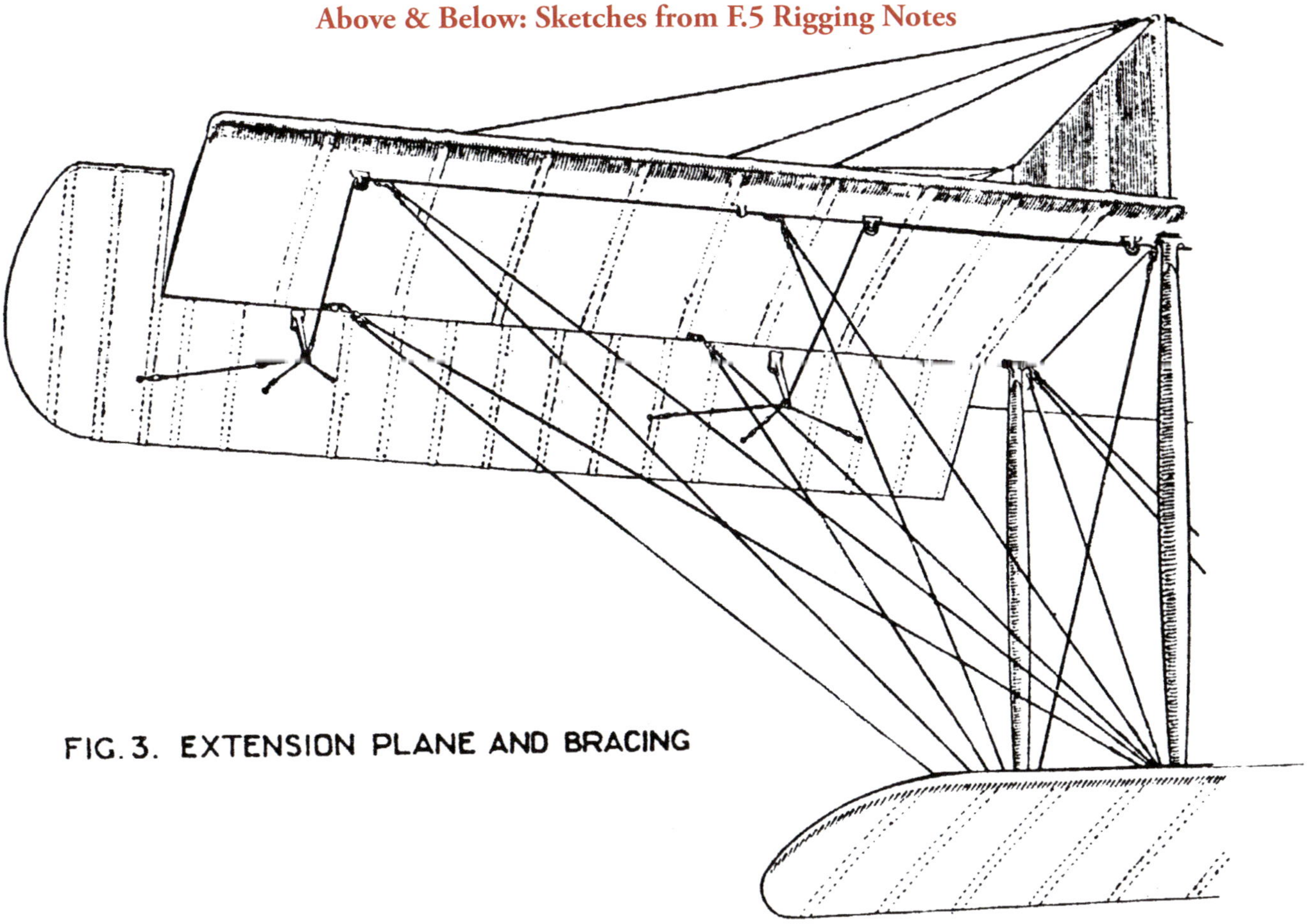

FIG. 3. EXTENSION PLANE AND BRACING

7. The Felixstowe Fury

Above: The hull of the Porte Super Baby (PSB).

Another type of flying boat, the largest flown to date, is the Felixstowe Triplane called the "Fury". It is built along the lines of the H-16 construction, having internal wiring with diagonal struts and longerons. ...In the first flight the boat left the water very easily and attained a speed of 94 knots (107 miles per hour) at sea level. No one was carried in the rear tail cockpit as this is to be washed out in favour of fighting cockpits on the wings. The empennage is being rebuilt along standard lines. The tail setting, in spite of the fact that, in the first flights was found to be neutral, is now being reset to two degrees incidence as originally intended. This will permit an overload in the tail. The landing speed was about 40 knots (46 miles per hour) which permitted the boat to land safely with no porpoising. The ailerons are balanced. Servo motors are attached to ailerons and rudder.

On another trial last Saturday (November 23, 1918) this boat flew with total load of 30,000 lbs, making about 9 lbs., per square foot loading. It attained a speed of 100 miles an hour. The landings were made very softly at 55 miles per hour.

Until the minor deficiencies of this boat are corrected and official trials made, no hard and fast conclusions are to be accepted. It has five 350 H.P. Rolls Royce engines with thrust lines in plane of middle wing. Two tractor and three pushers.[1]

This quotation is taken from a Memorandum by the USN on the latest types of British flying boats and describes the last of the series of Felixstowe flying boats, the largest designed by John Porte and his most ambitious – the Felixstowe Fury.

The Fury, was

designed as an experiment in seaworthiness.

The America *might serve for inland waters, the F boats were effective over the narrow seas when operating from a good harbour and well equipped base, but something was needed which could be landed in a heavy sea with safety and also rise again.*

It would need to be large so as to be little effected by moderate waves, strong to stand considerable rough usage, and exceptionally good at landing and getting off. It should be able to stay away from its base for days...

Getting off must be done rapidly, the long run usual, in smooth water could not be allowed and a "get off" in a series of leaps, would have to be reckoned on. To reduce the number of leaps, is

Above: Weighing the hull.

highly important. The loading laid down was 13 lbs/H.P., and not more than 8lbs/sq. ft.[2]

The 320th Meeting of the Progress and Allocation Committee on 10 April 1918, considered the work being carried out at the Felixstowe Air Station as follows:

Cmdr Porte is undertaking a design on the same lines as the N.4 but in order to get the initial experiment carried out as quickly as possible "he is pressing ahead with the design of a machine fitted with five R.R. Eagle VII engines in order that the experiment should not be delayed by waiting for the higher powered 800 H.P. BHP or R.R. engines."[3]

The primary object was to obtain a seaworthy flying boat that was capable of alighting and rise from the water under average North Sea conditions. With this end in view a basis of low weight per HP was taken and the question of military load, fuel, bombs, etc., was considered to be of secondary importance. It was estimated that the following load and performance would be obtained:

Crew of five.
Fuel 600 gals for 5½ hours at 93 knots full speed at sea level.
or 8 hours estimated at cruising speed.
Bombs 2 x 500-lb
Armament 10 Lewis guns and 650 rounds per gun.
W/T Pat. 54A[4]

Service ceiling was estimated to be 14,000 feet but Porte notes that this was more or less conjecture as there was not sufficient data for a triplane of this type to rely on.

The development of this large flying boat was recommended as it would provide private firms with assistance in developing large flying boats due to Felixstowe's previous experience in building and operating flying boats under actual service conditions. "As soon as the private firms are educated up to the Felixstowe standard it will be possible to give up Government designing and rely on the firms to continue the development."[5]

Porte had built a three-engined flying boat in 1916 that was known as the Porte Baby, so it was no surprise that the new project became the Porte Super Baby or P.S.B. in official reports.

In March 1918 Porte informed the RNAS Design branch that he had GA drawings of a new machine. On this letter the note "Is this the F.5 or a new design?" was answered with "new design." In reply the question was put as to the stressing of the machine, "Would it not be as well to get that checked over up here as soon as possible."[6] The P.S.B. was already well under construction at Felixstowe at this time as a request for five radiators, three large and two small, was sent the same month as they were "urgently needed in connection with the experimental Flying Boat now under construction."[7] The engine arrangement was for one pusher on the centre line and either side two engines back to back. It is interesting that this "machine has not received an official designation number, and

Above: This is thought to be the flotation testing of the Fury hull.

Right: Assembling the Fury

no requisition has as yet been received in Supplies for material or component parts."

It would seem that the P.S.B. was progressing under its own steam with officialdom being aware in a general sense but having a lack of detail knowledge of the design. Aircraft Production wrote to Porte, now a Lt Col, RAF, in April, in order to "remind you about letting me have some particulars of your scheme for the Super Baby, so that I can have an idea of the general lay-out of the scheme in its various stages. A few dates showing the stages of the programme would be of assistance."[8] On the 19th an official request was made for details of the flying boat "in order that official approval may be obtained for the machine to be constructed."[9] Approval was granted on 3 May 1918, under B.H. No.491, with the serial N123 allotted to the machine.

The fortnightly Reports of the Technical Department recorded the progress of the P.S.B. as follows:-

Felixstowe P.S.B. Boat Seaplane.
(Five Eagle VIII). No specification. Design and construction of one machine.

F/E 12.06.18.
Work on the hull and wings advancing well. Stress calculations are in hand.

F/E 26.06.18.
The hull is nearly ready to turn over. The wing fittings area being redesigned in accordance with recommendations of this Branch. Arrangements are being made to test a model of this hull in the National Froude Tank.

F/E 07.08.18.
No report has been received from O.C. Seaplane Experimental Station. Structural alterations have been asked for in the main planes.

F/E 21.08.18.
The engine bearers are erected and the centre engine is fitted in

Above & Below: The Fury as first rolled out with the early tailplane. The sheer size of the machine is well illustrated in these photographs.

Above: The Fury on the slipway at Felixstowe in front of the camouflaged hangars.

place. There is likely to be some delay in this machine from non delivery of engines and streamline wires. Tanks are fitted in the hull and internal fittings are proceeding. Wing root stay tubes have been fitted. The hull has been launched, the cradle and trolley are well advanced and the bulkheads have been tested.

F/E 04.09.18.

Streamline wires are not yet available, but all planes except top port have been erected with temporary wires. The tail fin is in position and the tail planes are being erected. Work is delayed by the non delivery of engines. A model has been forwarded for tests in the William Froude National Tank. Several structural alterations are required to bring the machine up to strength.

F/E 18.98.18.

Tests results of the hull model in the William Froude National Tank have been received by the N.P.L. Streamline wires are now being fitted. Internal and external fittings being completed. The tail bracing is complete. Several points of structural strength (are) *still outstanding.*

F/E 02.10.18.

N.123. As result of N.P.L. Tank Tests the forward step has been slightly modified. Main planes are being trued up. Collapsible dinghy has been commenced. All exhaust, oil and water pipes finished. Two jig engines are installed; the proper engines are urgently needed.

F/E 16.10.18.

Machine N123. Collapsible dinghy design complete. All engines in position and bolted down. Main filter and pumps installed. Truing up complete. Tail unit partially dismantled for strengthening up, and new tail structure now being designed. Petrol system being fitted including throttle controls. Flooring,

Above: As with all large aeroplanes of the day, the Fury had to be photographed alongside a smaller machine to illustrate its size. Here the chosen aircraft is a Sopwith 2F.1 Camel that, unusually, has a white outline to the underwing cockade. The Camel has a steel tube droppable landing gear.

float attachments and seats are in hand and main controls are all in hand. New outer wing struts being made. Modifications to being machine up to strength under examination.

F/E 30.10.18.
N.123. Erection is now complete and engines almost ready for trial run.

F/E 13.11.18.
N123. Ready for trial run. Structurally passed for experimental work.

F/E 11.11.18.
N123. New tail structure being made and Vickers pumps being fitted. Experimental turning indicators are also being fitted.

27.11.18.
N123. This machine has done successful preliminary flights. It is proposed to fit a larger tailplane and also to revise the petrol system using Vickers centrifugal pumps.

Period 12 to 31.12.18.
First time referred to as Felixstowe Fury.
N123. Being fitted with new tail structure and Vickers pumps. Second machine has been commenced.

Three weeks ending 01.01.19.
N123. Being fitted with new tail structure and Vickers pumps. Second machine has been commenced.

Month ending 31.01.19.
N123. Having Vickers pump fitted. New tail unit nearly ready for erection. New tanks in hand.
N130. Moulds and floors in preparation and keelson being framed up.

The reminiscences of Sqn Ldr Thedore D. *Pix* Hallam were published in 1919 and throw some light on how Porte was regarded by the men who flew his boats:

By this time Commander Porte had got out several experimental flying boats. He carried out his plans with a scratch collection of draughtsmen, few with any real knowledge of engineering; with boat-builders and carpenters he had trained himself; and he only obtained the necessary materials by masterly wangling. He frequently started a new boat and then asked the authorities for the grudged permission. But in all things connected with the building of flying boats his insight amounted to genius, and the different types of boats kept getting themselves born. His latest boat, known unofficially as the Porte Super Baby, or officially as the Felixstowe Fury, a huge triplane with a wing span of

Above: The object of mass admiration, the Fury at Felixstowe. Note the dazzle painted damaged hull in the foreground and the *Large America* dazzle painted boats in the background.

> *127 feet, a total lifting surface of 3100 square feet, a bottom of three layers of cedar and mahogany half an inch thick, and five engines giving 1800 horse-power, I flew successfully – it weighed a total of fifteen tons. On this test I carried twenty-four passengers, seven hours' fuel, and five thousand pounds of sand as a make - weight. Some idea of its huge size can be had when it is realised that its tail unit alone is as large as a modern single-seater scout.*[10]

It is evident that the RAF Technical Department, D Branch, on inheriting the P.S.B. from the Admiralty on the amalgamation of the RFC and RNAS to form the RAF, now wanted to control the experimental establishment at Felixstowe. It considered that Felixstowe should be handled exactly the same as a private firm and should follow the same procedure as detailed hereunder.

(1) Felixstowe is sent a RAF type specification for consideration and prepares GA Drawings for discussion, or would ask authority to proceed with an outline design likely to be of benefit to the RAF.
(2) On approval of the GA drawings, recommendations would be made to authorise construction. In this case there would be no contract.
(3) On approval being communicated to Felixstowe, construction could begin.
(4) Drawings and particulars required to criticise strength would be called for. Criticisms would be communicated to Felixstowe in order that they could be addressed by Felixstowe.
(5) Construction would be watched by this Branch.
(6) The machine would make test flights by a pilot under Felixstowe control but no other service pilot would be allowed to fly the machine until it had been passed for strength by this Branch.
(7) Test pilots of this Branch would fly the machine for criticism of its flying qualities.
(8) This Branch would give clearance certificate and detail tests that were required, and then official testing would be arranged.
(9) This Branch would remain responsible for the machine until the design had been passed for production or turned down.
(10) Within the limits of the above procedures no communication would take place with Felixstowe on the design, etc., unless through this Branch.[11]

It was in April that questions were raised about the allocation of five Rolls Royce engines to this project as "the boat is eventually intended for 3 Condor engines. Why not wait until the proper engines are available? We shall be very short of Eagle Rolls Royce engines soon, and there are many outstanding demands for them." In response J.G. Weir, the Controller of the Technical Department, stated that "the results to be obtained from this Large Seaplane are worth

Above: The Fury with its original tail takes to the water.

the allotment of Eagle 8 engines and if it has to wait for the Condor engines before first trial a delay of four months is almost inevitable."[12] With the reduced power the performance would suffer but this was accepted.

In June it was noted that the Felixstowe calculations and those in London had used different assumptions. In some cases the factor of safety was below 4 but they could be raised to 4 with minor changes.[13] It was understood that fittings would have to be redesigned "After the first machine, but on account of urgency they are only being strengthened with local additions at present."[14]

On 15 July the Technical Branch was informed that it was expected to commence erecting the machine within 10 days and the streamline wires, and the five Eagle VIII engines, all left hand tractors, were urgently required. On 8 August Felixstowe wrote that the lack of streamline wire would soon bring the erection of the machine to a halt. It was requested that as Messrs Brunton should have at least some of the wires ready, that they be dispatched by train.[15]

On 14 August 1918, Porte was notified that the P.S.B. Boat Seaplane had been officially registered as the Felixstowe Fury and all "future questions concerning machines of this design should be referred to by these names."[16] Despite this the term P.S.B. still appears on Felixstowe documents. This same month a model of the Fury's hull was sent to the William Froude National Tank for testing and as "the fact that the machine itself is nearing completion" it was requested that tests be carried out as a priority.[17]

The Fury was the largest British aircraft of its day. The hull had been developed from the line of Felixstowe flying boats and was similar in construction. It is considered that the Fury also drew on the lessons learned from the massive Curtiss Model T triplane flying boat that had been assembled and flown at Felixstowe.

The hull was 60 feet in length and built on the Porte principle that followed landplane practice with four longerons built into a cross braced box girder, but more minutely designed in detail, the wiring being more systematic and all struts and fittings carefully planned. The keelson and floors were built up of mahogany in a lattice girder construction instead of being solid, as on the smaller boats. Very strong bracing was provided and special struts to transmit landing loads to the boat bottom. The hull was planked diagonally with two skins of cedar. The boat bottom forward of the step had an additional layer of planking making three skins in all. It was laid fore and aft and separated from the second skin by a layer of glued calico. This same material was used between the other two skins. Instead of the top and bottom of the fins meeting at an angle, the diagonal planking was steamed and bent around the chines and fin tops to form a curve of radius 5 to 8 inches. This eliminated the problems that occurred with the earlier F-Boats at these joints, while the thick edge gave a great increase in displacement for a small alteration of draft and made heavy overloading possible.[18] The bottom plank did not end at the fin edge but was carried right over the rounded fin edge and, with another bend at the waist, up to the upper

Above: The Fury with its original tail takes to the water.

longeron. This provided a very flexible fin while avoiding the risk of leakage that characterised the built up form of fin edge. "It required very careful work to bend and fit the planks, but when completed and secured with the strong rubbing strip, it was a sound job." The hull bottom was slightly concave. The hydroplaning efficiency of this hull was little inferior to that of the F.5. Bulkheads of Willesdon canvas were provided and tested by filling the fully separate compartments with water.

The upper and middle wings were of equal span of 123 feet, with the lower wing being shorter with a span of 89 feet.[19] In order to avoid deep wing floats there was no dihedral on the bottom plane, the other two were given 2°. Balanced ailerons were fitted to the top and middle wings only and were coupled together. Control wires ran inside the wings. The centre section was separate from the wing bracing being supported by the struts.

The tailplane resembled that of the Curtiss Model T being of biplane form with a small balanced rudder mounted wholly above the tailplane, and twin unbalanced rudders mounted between the tailplanes. As with the Model T the lower plane could be adjusted in its entirety in the air for fore and aft balance. This was operated by a lever beside the pilot, through a screw attached to the leading edge of the lower tailplane. The screw was raised or lowered by a nut driven by a chain and sprocket. On the upper tailplane the elevator was worked in the usual way. The machine proved to be tail heavy in a glide and the tail was replaced by a larger more conventional biplane unit with triple rudders, a fixed lower plane with an elevator that could be adjusted by a lever at the pilot's side to correct for fore and aft balance.

An unusual feature in the building of this boat was that the hull was constructed upside down, which method gave the men access to their work, and allowed proper inspection.

Another new departure was the launching of the hull alone. Previously, machines had been completely fitted up before being placed in the water but then leakages might be discovered in inaccessible places. With this boat, the hull was tested for watertightness, while the inside was still clear of obstructions, and all leaks stopped.

It had been hoped that 600-hp Rolls Royce or B.H.P. engine would have been available for this boat but their unavailability meant that five Rolls Royce Eagle VIII engines were fitted instead. It was estimated that the arrangement selected, two sets of tandem mounted engines and one pusher, would allow for 67% of propeller efficiency, or a loss of 8% from the 75% allowed for a well designed single airscrew. This meant a loss of 150-hp out of the available 1,800-hp. "The loss at full speed is probably about 6%, but the loss at hump speed is a great deal which is the real crux."

Six fuel tanks with a total capacity of 750 gallons were provided in the hull. They were all placed horizontally with a deck over them leaving plenty of room between them and the hood for access. "Recently the tank capacity has been doubled, making it 1,500 gallons or 4¾ tons."[20]

Cooper Servo-motors were fitted to all controls but were to prove not to be needed as the Fury proved remarkably light on the controls.[21] Major Arthur Cooper had been asked by Porte to design an entirely mechanical servo-motor for the Fury to replace the Sperry servo-motors that had proved troublesome. A lever could lock these servo-motors out of gear when not required. Cooper also assisted in the testing of the Fury as related hereunder.

Above: The Fury; its hull was the best developed by Porte and threw up less water than previous hulls.

An adjustable tailplane presents some difficulty in the case of a biplane tail plane. In the case of the P.S.B. the elevator on the lower plane was operated separately by a long lever on a quadrant centred on the elevator control shaft. This was used to obtain trim and the upper elevator for extra control in the usual way. The result was quite satisfactory, but owing to hull interference the efficiency of the lower plane was relatively low.[22]

Engine run ups were conducted on 30 October for final adjustment with one day being set aside for final inspection, launching being set for the 1 November. The launch was delayed until the 12 November 1918.[23] The Fury taxied around the harbour and was taken just off the water several times on three engines and found to behave quite well. On the next day

She was taken off by Col. Porte and Major Cooper with 9 passengers and 300 gals. Fuel i.e. a useful load of about 4,000 lbs. All five engines were running but the centre one was not opened out except for a top speed trial on one flight. The machine was taken off and landed four times, three of which were straight, but on the other flight turns were done with about a 15° bank.

The machine got off and landed well, was fairly clean on the water but at one speed the wash met over the tail of the boat though it did not affect the tail planes.

The machine was rather tail heavy in the air due to an error in the setting of the tail plane but was well under control both with the engines on and off. It was intended to alter the tail setting and continue trials as soon as possible.

Major Cooper considered more tail area was needed but was satisfied that the machine was safe to carry out the trials as she stood. The high centre of thrust did not affect the stability as had been anticipated.[24]

Major John D. Rennie had worked with Porte at the Felixstowe Experimental Section and was a defender of Porte's method of construction.[25] In his Royal Aeronautical Society address he noted that with

regard to fin and rudders, these were fairly large in the Felixstowe F boats, but barely sufficient for control with one engine cut out completely. The P.S.B. was very satisfactory in this respect due to the engine arrangement of three propellers abreast and three fins and rudders in their slipstreams.

The rudder, ailerons, but not the elevators were balanced; servo-motors were also fitted to all control surfaces as it was thought that while adequate control under certain flying conditions might not be within the capability of the pilot, there being no previous experience with a boat of that size. However she proved to be extraordinarily light on all the controls and superior to the much smaller "F" boats. The servo-motors were removed as Col. Porte decided that any advantage to be gained did not warrant

Above & Facing Page: The revised tailplane of the Fury was a much neater arrangement. Note that the cockades were applied to the bottom of the middle wing.

the additional weight and complication.[26]

The 14th of November saw a test to ascertain the maximum air speed attainable. The pilots were Major Cooper and Capt Scott. A large circuit was made over the Cork Lightship whilst climbing. The speed runs from the Felixstowe Pier head to Walton Pier Head and vice versa were accomplished without incident. Average speed was 92.25 knots (about 114 mph). After the conclusion of the speed runs a portion of the port tractor engine's exhaust pipe broke away, exposing the radiator and a strut to the hot exhaust. Leakage of the radiator occurred prompting a landing. Whilst flying the servo motor serving the rudder jammed and could not be disengaged. Cooper remembered that one of the new servo-motors that he had designed

had been fitted to the rudder control, and I decided to try it out. When I switched on the servo-motor the rudder became fixed and could not be moved in either direction. I could not disengage the motor by the lever owing to the load. I had prepared to land without rudder when at last I managed to disengage it. The machine took to the water very well. The cause of the jamming of the control was that the wire from the motor had been connected to the wrong wire from the hand control so that it pulled against it instead of assisting.[27]

This would have been one of the longest flights the Fury had made. Most trial flights were of a couple of minutes only.

The water performance of the machine was far ahead of expectations. At a gross weight of 2150 lbs. she got off with three engines only corresponding to 24 lb/H.P.

There was no sign of diving at low speed porpoising. The nose may have been too high for least head resistance, when hydroplaning, but there was not the least difficulty in getting off. Landing was extraordinarily easy and shockless.

From preliminary testing the following points were brought out:

The speed full out was 92.5 knots with a weight of 25,650 lbs.

The hull was described as "staunch and tight."

The boat was flown without the servo motors.

The machine was tail heavy in a glide with the fixed tail set at 2°. A new tail unit of greater area was designed.[28]

The petrol pumps were not entirely satisfactory and it was proposed to fit Vickers centrifugal pumps.

No trouble was experienced in getting the boat in and out in a light wind.

The machine handled easily on the concrete areas by about 20 men. The boat handled very easily on the water.

"The trials showed no important defects and the results are considered very promising."[29]

On 23 November 1918, a test was undertaken to determine what load the "P.S.B." could take off smooth water and without wind. The pilots were Majors T.D. Hallam and Ronald Moon.[30] Short straight flights were made with the moveable tailplane set at neutral. The fixed tailplane was set at 2°. After alighting the moveable tailplane was set at 1° down. A second and third flight was made at about a height of 40 feet. During the fourth flight the boat got off in 30 seconds and several turns were made at an altitude of 1,500 feet. The load was 30,450 lbs. Takeoff speed was 55 mph and landing speed 60 mph.

For the second test the weight was 32,500 lbs, and the takeoff speed was 62 mph after a run of approximately 49 seconds and after climbing to 100 feet, the boat was landed at 60 mph.[31] "The test showed, that the resistance of the steep V bottom was not excessive. The loading was 18lbs/H.P."[32]

November also saw tests on the engines to determine that all were running up to speed and that the propellers were performing satisfactorily. The tandem engine installation was new and not fully understood. The USN had a similar problem with the NC flying boat engine configuration for the trans-Atlantic flight of these machines. A collapsible lifeboat was made at this time for the Fury for the "safety and convenience of the crew." A report of late December noted that the boat was being fitted with a "new tail structure and Vickers pumps."[33]

Grain test report NM250 is not dated but the speed and climbing trials recorded therein are dated 18 and 19 June 1919. The triplane was described as stable longitudinally at all speeds by setting the adjustable auxiliary elevator. It was stable laterally and directionally.

Under the heading "Controllability," longitudinally the "Auxiliary elevator makes machine very nice to fly, but main elevator is slightly soft at low speeds especially when landing." Lateral controllability was very good; directionally it was very good "but heavy, should be fitted with servo motor." The cockpit layout for controls was good with the only suggestion being made that the engine altitude control should be fitted in reach of the pilot. The machine was not tiring to fly except when manoeuvring when a servo motor was "needed for rudder control." The boat alighted "very easily and gently but elevator control is slightly soft at low speeds."[34]

Little information seems to be available on the proposed armament of the Fury but one source suggested that the armament was four Lewis guns and a heavy bomb load.[35] Knowing the number of guns carried by the F.2A this seems too small a number for a boat the size of the Fury. In one file there is a suggestion that the "fighting tops" developed with the F-Boats could be used on the Fury, but the memo corrects the questioner and confirms that the "Fighting Tops" were erected on an F.5 boat, not the Fury. The N.4 flying boats were considered for the mounting of a C.O.W. gun, and although

Above: With an all white hull and interplane struts and fabric surfaces probably in P.C.10, the Fury is ready for its record flight attempt.

an interim design between the N.3 and N.4 specifications, it appears that this Porte boat was considered for mounting this weapon. The C.O.W. Gun weighed 200 lbs and each complete round weighed 2¾ lbs with the mounting and sight adding a further 30 lbs. With respect to the P.S.B. Porte was notified that designs were in hand for an incendiary shell, also an HE shell with a sensitive fabric fuses. An armour piercing shell had been designed and supplies were being made of an HE shell with a "fairly sensitive percussion fuse." The gun could fire single shells or six automatically at a rate of about 120 rounds per minute. It was suggested that an officer of the Technical Department be sent to Felixstowe to discuss the armament of the P.S.B. in order to keep in touch with the "latest developments."[36]

From the testing of the Fury it was determined that the *extraordinary behaviour of the hull in rough seas was due mainly to the buoyancy of the hull and the lines adopted. With a load of 28,000lbs, the chine at main step was not submerged. Under all conditions, the propellers, cockpit and tail plane were clear of water thrown up. From this point of view it was found that the bow sections could be improved considerably. These were too blunt, resulting in unnecessary pounding when getting off or landing in a rough sea, or at moorings. It was decided in any future designs to drop the keel profile forward, keeping everything else the same, thus obtaining a finer entry, without loss of buoyancy forward.*[37]

The development work on the Fury appears to have proceeded at a more leisurely pace with the coming of peace. It was proposed that the Fury undertake a trans-Atlantic flight as it was well within its range. On 28 November 1918, the Assistant Controller, Design, noted that the Fury would not fly again until the New Year owing to changes being made in the tail structure. A second machine to undertake the Atlantic Flight was required "as a demonstration and for encouragement of commercial aeronautics." As there would be great difficulties in having this built by private firms as the design and drawings were in an incomplete state, the machine should be constructed at Felixstowe and instructions had been issued to Lt Col Hallam to go ahead.[38]

In a report of 21 December 1918, it was stated that construction of the second machine had been commenced at Felixstowe almost entirely from material drawn from RAF stores. For the few small parts that were required it was asked that they be given 1st class priority, due to the urgency of completing the machine for the Trans-Atlantic flight. This was to be made from Cape Broyle Harbour, Newfoundland. The Navy was requested to make shipping available to transport the flying boat to here and for ships to be along the route. The range of the boat was estimated by Col Porte as 1400 statute miles "Although it may be possible to get another 100 out of her." The Report asked "Is it or is it not worth while that the first Aircraft to cross the Atlantic shall be built, designed,

Above: Approaching the wreck of the Fury.

and manned by the British."[39] In the 27 March edition of *Flight* it was stated that the "Air Ministry is hurrying on the preparations for an attempt by one of the "Felixstowe Fury" flying-boats, which would be navigated by Col. J. C. Porte, R.A.F. The attempt would be made from Newfoundland with stops at the Azores and possibly Lisbon."[40] There was a change in direction in June as the prototype was to be used for the flight. It was to be dismantled for shipment to the US for the attempt. It is assumed that the time factors would not allow for the second machine to be completed in time with the number of contestants already in the race. The success of the Alcock and Brown Vickers Vimy doomed this project, although it has been suggested that it was cancelled due to a lack of funds.

The Felixstowe Fury story ended on 11 August 1919, when the machine crashed at Felixstowe. The flying boat was to have to flown to South Africa via Gibraltar, Egypt and the great African lakes. The machine was provisioned and equipped for the trip. The crew comprised Lt Col P.F.M. Fellowes as OIC; Major Edwin Roland Moon, DSO, pilot; Capt C.L. Scott, DSC, pilot; 2/Lt J.F. Arnold; Chief Engineers W/O H.S. Locker and W/O J.G. Coburn; Lt S.E.S. McLeod W/T operator. The Fury was taking off for Portsmouth for the start of the African flight. It had just rose from the water when the port wing dipped, struck the water and the machine crashed in Harwich Harbour.[41] McLeod was drowned, the rest of the crew surviving. It appears that the machine lifted off before it reached safe flying speed. Both Porte and Rennie had left the service before this flight and there was no technical officer at Felixstowe to supervise experimental flying.

What follows is an eyewitness report of the tragedy. The copy in the Felixstowe file is not signed and the initials appear to be RDO. The author of this report wrote to Porte on 12 August that he and General Ellington went up to Felixstowe and "saw it all."

He taxied the machine in a southerly direction first, I.E. to get clear of the jetties, buoys, etc., and then turned left-handed towards the mouth of the Orwell with the object of course, of getting down wind before starting to rise. It struck me that the machine was rather left wing down while taxing and the left float was frequently in the water; this must have been due to the machine turning rather left-handed most of the time. It taxied about ¾ mile down the river and appeared to be perfectly under control on the water. It then turned round heading rather east of N.E. Moon then opened up the engines and taxied the machine for about 500 yards; the machine seemed to me to take a very long time to rise out of the water at all and appears to be practically as low in the water the whole time as it was when standing still. After about 500 yds, however, the machine began to skim along the top of the water and then rise about 5 ft. in the air for about 20 or 30 yards, then dropped down on to the water again keeping level the whole time, i.e. the tail of the machine was not at all down when it got off the water nor did the nose of the machine drop when it dropped back on to the water again.

Above: The wreck of the Fury after recovery.

I then had a look at the wind and saw that it had changed further round to the north so that the machine was now taxing at an angle of about 45° to the wind, the wind of course coming from the right side of the machine. The machine went on taxing for about another 100 yds. or so and then rose from the water again very awkwardly. It turned right-handed while in the air so as to be heading momentarily almost straight into the wind. It never seemed, however, to have flying speed and a few moments afterwards it dropped towards the water, turned left-handed, the port plane struck the water, and the machine crashed completely. The total time from the moment that Moon opened out his engines until the crash occurred was 59 secs, and it would seem that each time the machine actually rose from the water it was doing nearly 60 m.p.h. According to what Moon stated the machine was very nose heavy. He had the moveable tail plane and elevator right back but even then could not get the nose up.

I had a look at the fixed tail plane when the machine was brought in and it certainly was up, I think in its maximum position, so that there was no question of the controls, at any rate these controls, being reversed.

The measurements of the rigging appear to be absolutely correct and the machine had apparently flown twice since its accident on test some weeks ago.

The only reason that I can see to account for the accident is the loading of the machine. It was not fully loaded, apparently about 2,000 lbs. short of the full load, but I fancy that the load was distributed wrongly. I spoke to one of the mechanics who was responsible for the petrol; he stated that various spares had been loaded in the back part of the machine and that in order to counteract these he had left the rear tanks empty and filled the front tanks about 2/3rds full, the centre tanks being pretty well filled up. No attempt seems to have been made to weigh out the various spares that were carried and one knows that aeroplane spares look much heavier than they are. However in addition to this there was considerable weight of wireless apparatus, some rifles, ammunition, rations, thermos flasks, officers kits, and other oddments, most of which seem to have been put in more or less at the last moment and in the front part of the machine…

I see according to the Evening papers that the hull of the body opened and let the wireless operator – Macleod – drop to the bottom, this is quite wrong: the port side of the hull did open out more or less along the water line but Macleod was still inside when the machine was dragged to the beach. His position was apparently down below so he really never had a chance.[42]

Flight reported on 14 August 1919, that the Fury made towards Landguard Point and then turned to get a favourable wind but there was apparently some difficulty in clearing some buoys and a river boat.

At the inquest on 12 August 1919, Lt S.E.S. MacLeod who was killed in the crash of the Fury received a verdict of

Fury Specifications				
Source	Development of Flying Boats by Maj W.F. Vernon	Some Notes on Flying Boats. RAeS Journal 1921	JM Bruce Data	Grain Report NM250
Span – Upper Two	123 ft		123 ft	123 ft 0 in
Span Lower	89 ft			119 ft 0 in
Length Overall			63 ft 2 in	63 ft 0 in
Length Hull	60 ft	60 ft		
Height			27 ft 6 in	27 ft 6 in
Max Beam		12 ft 6 in	12 ft 6 in	
Chord	10 ft		10 ft	10 ft 0 in
Gap	8 ft		8 ft 6 in	
Stagger			Nil	
Dihedral – Upper Two Wings	2°		2°	
Dihedral Lower Wing	Nil		Nil	
Incidence				4½°
Incidence Tail				2°
Areas in Sq. Ft.				
Top Plane				1,240
Middle				1,080
Bottom				840
Wings		3,108	3,108	3,160
Ailerons (Top)				148
Ailerons (Middle)				148
Tailplane (Top)				168
Tailplane (Bottom)				102
Elevators (Top)				66
Elevators (Bottom)				40
Tailplane & Elevators			378	
Fin			86	
Rudder			58	
Gross Weight, Lbs.		32,000		21,883 (light) 25,253 (medium)
Climb (Light Load/Medium Load)				
To 1,000 Ft				1 min 25 sec/1 min 50 sec
To 3,000 Ft				4 min 20 sec/5 min 30 sec
To 5,000 Ft				7 min 25 sec/10 min
To 10,000 Ft				19 min 05 sec/28 min 20 sec
To 13,000 Ft				31 min 10 sec/—

"accidental death" – no light was thrown on the cause of the accident.[43] Charles Gates "served the last four months of my RAF Service there and saw something of it (*the Fury*). I was friendly with its wireless operator, Lieutenant Sam McLeod *(sic)* who was drowned when it crashed on 11 August 1919 when taking off on attempt to fly to Cape Town...McLeod's station was in the bowels of the ship amongst the fuel tanks. I had often urged him not to be there on take-off and landing. That crash was some months after I had been sacked with a 'gratuity' of Ninety Pounds for four years war service. I never flew in the Fury but, with McLeod, tested its radio equipment in a F2A."[44]

Major Rennie in his RAeS paper made the following observations.

With regard to the "Fury," it is quite obvious many important facts are not generally known. Prior to the accident she had done about 30 hours flying, and in every way was found most satisfactory. Under the control of expert boat pilots, such as Colonel Porte, Major Hallam, Hobbs, Wright, and Cooper, porpoising to any serious extent was absent when taking off. The tendency was there, but sufficient elevator control was available to keep it down. Owing to the fact that the tail plane being covered by the propeller slipstream, there is more elevator control over the range of speed between the hump (speed) and taking off, than is probably realised.

The C.G. position had been determined by weighing in the usual way, contrary to what is stated in the report on the accident.

Several months before the accident Colonel Porte and myself were demobilised, there was therefore no technical officer in charge. No one knew where the C.G. ought to be or took the trouble to find out. The result was that the boat was loaded up with spares, etc, and the fuel, of which there was tankage for 1,500 gallons, most likely distributed in the tanks such that the final C.G. position was at least .5 of the chord. Further, from a reliable and intelligent member of the crew, the late Major Moon attempted to take her off, as he had done on a previous occasion, before the minimum safe flying speed had been reached. As loaded, she was underpowered, there was therefore little available h.p. for acceleration, once clear of the water with the inevitable result."[45]

Rennie was of the opinion that the Fury had been taken off in a stalled condition. He recorded that the "elevator control will be sufficient to trim the boat through nearly the whole range of attitude" maintaining the wings at a safe flying angle. Unless there was ample power for acceleration or "take off" speed, say 5 knots above stalling speed and elevator corrected, "there is the possibility of the boat leaving the water stalled."[46] Work on the second Fury, N130, was abandoned in September 1919 after the loss of the first machine. It is not known if the improvements mentioned by Rennie were incorporated into the second machine.

After he left the RAF Porte had been employed by the Gosport Aviation Co Ltd and had undertaken a number of designs of commercial flying boats for that concern. These were based on the work he had undertaken at Felixstowe and the Gosport G.9 triplane, powered by three 600 hp Rolls Royce Condor engines was obviously based on the Fury. Porte died in October and the company folded when post-war orders did not materialise.[47]

Rennie recorded that from every point of view the Fury was "the best design turned out at Felixstowe. It was found that the normal load could be increased to 28,000 lbs, under which loading, seaworthiness, ease of taking off and launching were superior to that of previous F-boats. Loading tests continued up to 33,000 lbs, at which Colonel Porte took her off in Harwich Harbour. Landing under all loads was without perceptible shock." According to Rennie the extraordinary behaviour of the hull in rough seas was due mainly to the buoyancy of the hull and lines adopted. With a load of 28,000 lbs the chine at the main step was not submerged and under all conditions, the airscrews, tailplane and cockpit was clear of water thrown up. The bow sections were too blunt resulting in unnecessary pounding when getting off or alighting in a rough sea or at moorings. It had been decided that in any future designs to drop the keel profile forward but keep everything else the same, thus obtaining a finer entry without loss of buoyancy. He lamented that the hull improvements incorporated in the design of the Fury's hull had not been followed up as "these improvements led to a hull greatly superior to that of the "F" boat hulls, it is most unfortunate that the development of this type of hull has not been proceeded with, more especially when one realises the achievements and performance of these boats."[48]

Chapter 7 Endnotes

1. Memorandum for Captain Schofiels, 28.11.1918. -pencilled on copy "For Commander Hunsaker" Copy in NAHC
2. Maj W.F. Vernon, Report: *Flying Boat Development,* revised by Col Porte. Copy in TNA FILE MUN8/35.
3. Memo: *The development of the Large Flying Boat from the present types of Felixstowe Flying Boats now in production is being carried out in roughly four stages.* 08.03.18. TNA AIR 2/989 & TNA File AB/286/154/5 Part II. A Form C – *Request for Requisition to build Experimental Aircraft for the General benefit of Aeronautics,* was lodged on 06.06.18, for one machine. This Form stated that the P.S.B. was between the N.3 and N.4 flying boat specifications. TNA AIR2/987.
4. Statistics from Memo by F.W.T. Hewlett for Controller Technical Dept. 03.05.18. TNA AIR 2/989.
5. Memo: *The development of the Large Flying Boat from the present types of Felixstowe Flying Boats now in production is being carried out in roughly four stages.* 08.03.18. TNA AIR 2/989.
6. Wing Cmdr Ogilvie, Assistant Controller Design, to Porte, 28.03.18. TNA AIR 2/989.
7. Porte to Controller of Tech Dept, DS Section. *Radiators.* 28.03.18. TNA AIR 2/989.
8. Design Branch, Aircraft Production to Porte. 17.04.18. TNA AIR 2/989.
9. F.E.T. Hewlett for Controller, Technical Dept, Aircraft Production to CO, RNAS Felixstowe. 19.04.18. Note that the RAF had been officially formed on the 1 April but the old nomenclature is still in use.
10. PIX (Theodore D Hallam) *The Spider Web,* Willliam Blackwood & Sons, 1919.
11. A-CTD Design to D/C TD. 10.05.18. TNA AIR 2/989.
12. Copy of File notes. 22.04.18. TNA AIR 2/989.
13. "P.S.B. Stresses." Technical Dept, Design, to OC, Felixstowe. 19.06.18. TNA AIR2/987.
14. *Ibid.* "P.S.B. stresses and factors of safety." 11.07.1918.
15. OC Felixstowe Experimental Station, to Controller, Technical Dept, Dept of Aeronautical Supplies. 08.08.1918. TNA AIR 2/989.
16. Assistant Controller, Design to OC, RAF Seaplane Experimental Station, Felixstowe. *Standard Nomenclature of Aircraft.* 14.08.18. TNA AIR 2/989.
17. A.J. Miley for Assistant Controller, Design. 29.08.18. TNA FILE AIR 2/986 Mr G.S. Baker of the NPL pointed out that it was not until after the Fury was built that the NPL was asked to carry out tank tests on a model of the hull, and these had predicted that the machine would porpoise at 27 knots. This actually proved the case. Baker was replying to Rennie's paper. *RAeS Journal* 1921, P.123.
18. Mr Ward of the US Naval Aircraft Factory considered that "all experience at Felixstowe tended to show the necessity of transverse bracing in the hull. What was wanted was resiliency with lightness "and we obtain this with the use of ribs. When ribs are left out, the boat becomes too rigid, and in handling on trucks, the rigid bottom will fail. All experience at Felixstowe tends to show this. In our latest design of the Porte Super-Boat, comprising a hull 68' feet long, Mr. Porte reinforced this boat with continuous ribs from the upper longerons to upper longeron. The fin edges on this boat are circular, having a 8" radius in order to allow the ribs to be bent around the fin edge. This provided resiliency when the boat lands on the water, and when it is being hauled on the handling truck." R.E. Bassler to J. Hunsaker RE: Visit to Philadelphia Naval Aircraft Factory. No date. See accompanying diagram of the hull of the Fury.
19. See Specification Chart at end of chapter.
20. The quotes in this section are from - Maj W.F. Vernon, Report: *Flying Boat Development,* revised by Col Porte. Copy in TNA FILE MUN8/35.
21. Jack Bruce made the following note after an interview with Maj J.D. Rennie on 15.09.1964:- "Fury" Aileron servo device consisted of wind-driven gadget on central strut. Cables ran up this strut and passed over a pulley. When load on cable exceeded a certain value it pulled the spindle to one side and cut in the servo. Letter from JMB to W. Gunston, 13.11.1989. Copy in RAFM J.M. Bruce Collection Box 44.
22. Rennie, J.D. "Some Notes of Flying Boats", *Royal Aeronautical Society Journal,* 1921, P.123.
23. Sturtivant & Page give 11.11.1918 as date of first flight.
24. Inspector's Report: *Visit to RAF Station Felixstowe by Maj Bumpus, RAF.* 13.11.1918. TNA AIR 2/989.
25. Rennie was to design the Blackburn Iris flying boat.
26. Rennie, J.D. "Some Notes of Flying Boats", *Royal Aeronautical Society Journal,* 1921,

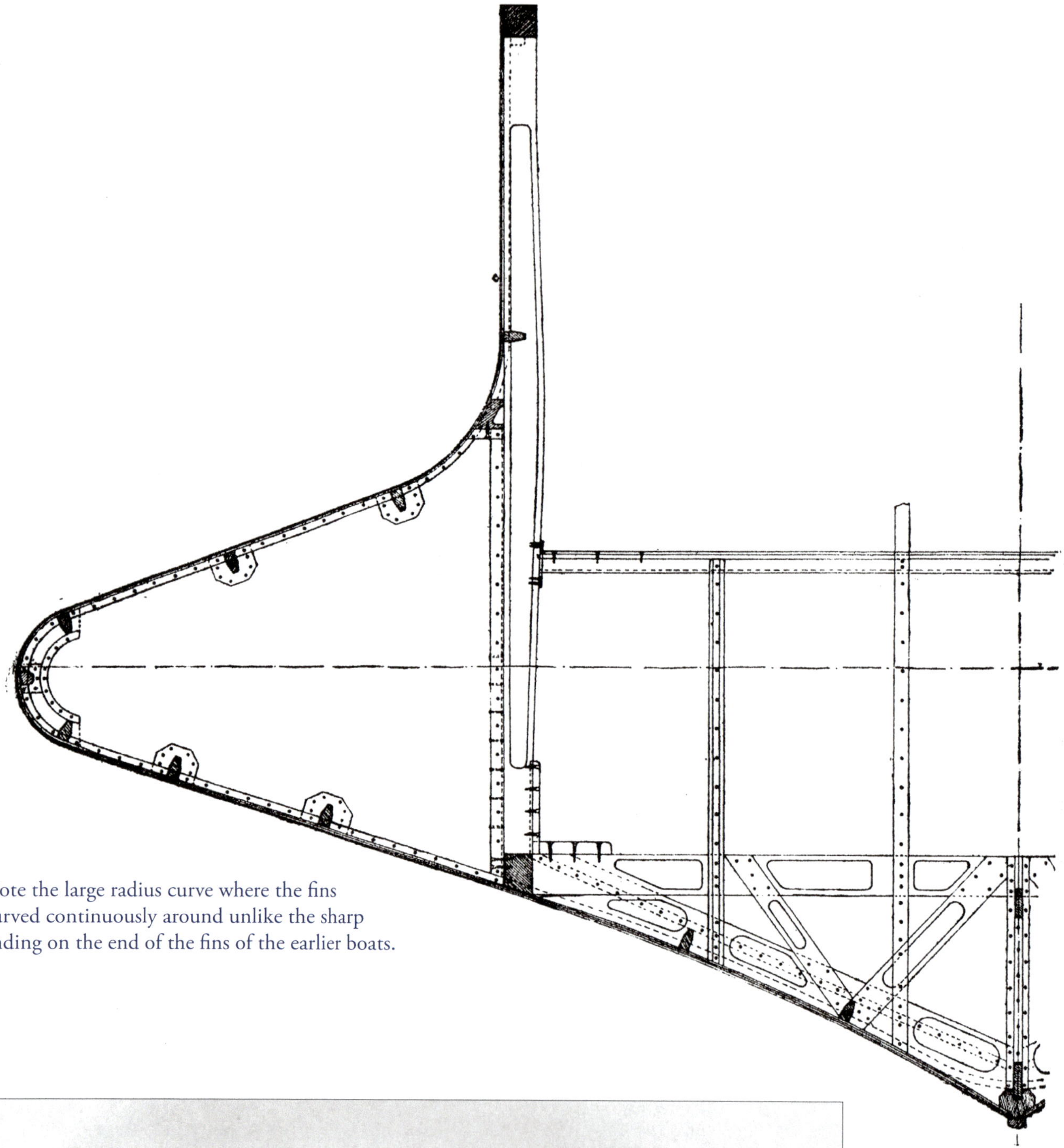

Note the large radius curve where the fins curved continuously around unlike the sharp ending on the end of the fins of the earlier boats.

Left: Wreck of the Fury.

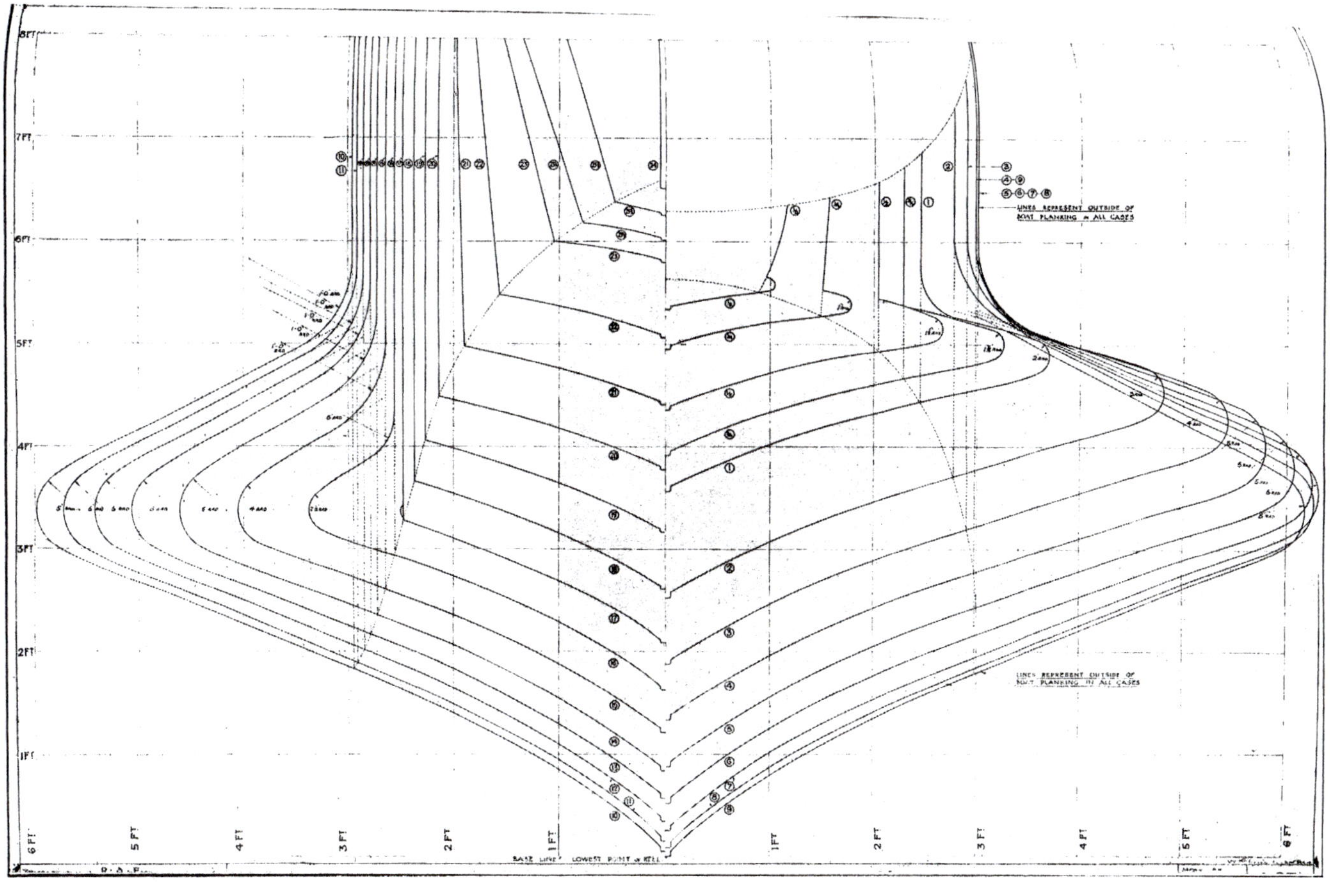
8FT
7FT
6FT
5FT
4FT
3FT
2FT
1FT
LINES REPRESENT OUTSIDE OF BOAT PLANKING IN ALL CASES
LINES REPRESENT OUTSIDE OF BOAT PLANKING IN ALL CASES
BASE LINE LOWEST POINT OF KEEL
6 FT
5 FT
4 FT
3 FT
2 FT
1 FT
1 FT
2 FT
3 FT
4 FT
5 FT
6 FT

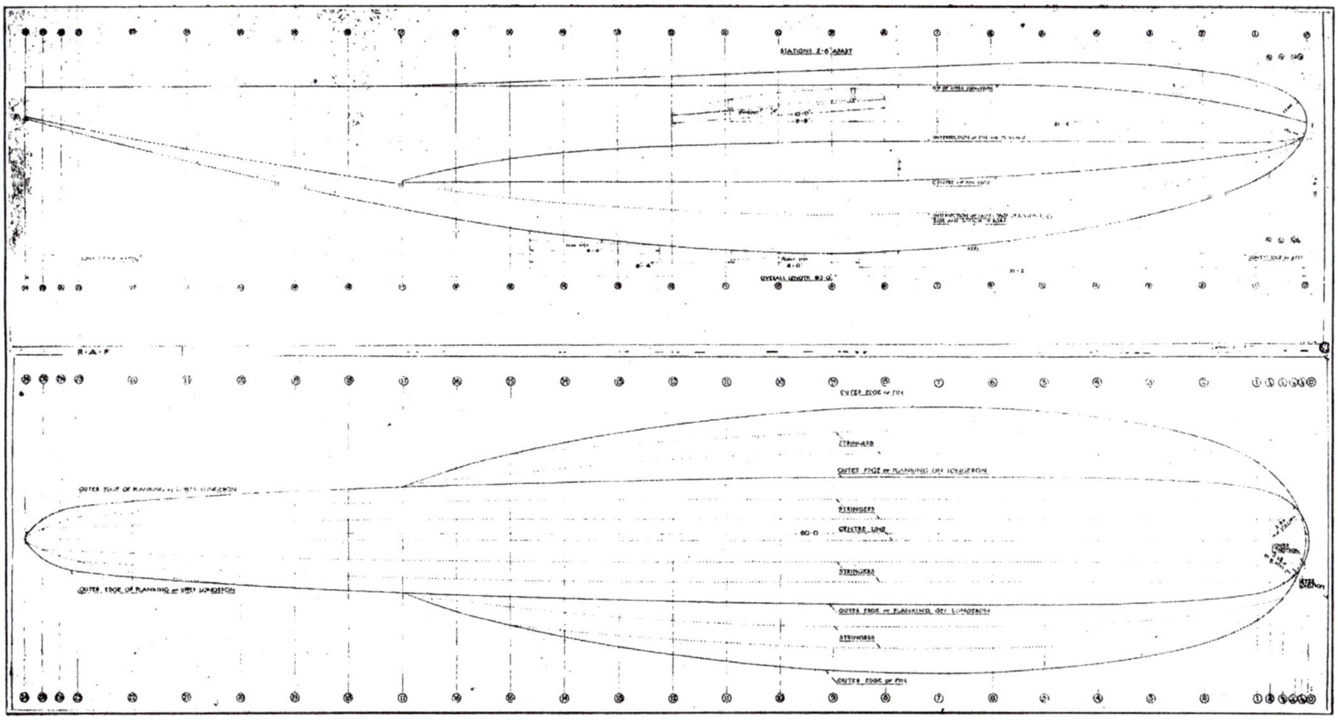
STATIONS 2'-6" APART
OVERALL LENGTH 60'-0"
R.A.F
STRINGERS
OUTER EDGE OF PLANKING ON LONGERON
STRINGERS
CENTRE LINE
STRINGERS
OUTER EDGE OF PLANKING ON LONGERON
STRINGERS

P.123.

27. Hadingham, E. *The Fighting Triplanes,* Hamish Hamilton, UK, 1968, P.158.

28. According to the article *The Felixstowe Fury or Porte's Super Baby,* in the series "Looking Back", (Journal unknown), the whole of the tail structure had to be dismantled as a result of stress calculations. "Designs were put in hand for a new tail unit but because of the delay in proving the structure as a whole, the old unit, with strengthening, was refitted and the Fury flew in November 1918." Copy article in J.M. Bruce Collection Box 22, RAF Museum.

29. Memo: Felixstowe "Fury" & Trial Reports Nos. 15 to 17. TNA File AIR2/986.

30. Moon was the pilot when the Fury crashed killing one crew member. Unfortunately Moon was to be killed in a F.5 crash at Felixstowe on 29 April 1920. The F.5 was on an instructional flight when Moon took over the controls as he had to do a glide. The witness stated that when about 1,500 feet the machine received a bump on the tail that threw the machine out of control. It developed a spin and despite Moon's attempts, the distance from the water did not allow for a full recovery. Moon and Lt Albert John Fyfield were killed, while Lt Fonsepca, a Portuguese officer, and Mechanic Bass had not been seen since the accident, and were presumed dead. "Inquest on Felixstowe Victims." *Aeronautics*, 06.05.1920. Quoted in J.M. Bruce Collection, Box 22, RAFM.

31. Trial Report No.19. P.S.B. – Preliminary Trials. TNA File AIR 2/986.

32. Maj WF Vernon, Report: *Flying Boat Development,* revised by Col Porte. Copy in TNA FILE MUN8/35.

33. Seventieth Report of the Controller, Technical Dept., for the three weeks ending 1st January 1919. Copy in J.M. Bruce Collection Box 25, RAFM.

34. TNA AIR1/1204/204/5/2616.

35. Bruce, JM. *British Aeroplanes 1914-1918,* Putnam, UK, 1957, P.252.

36. Letter: Technical Department D (S), Aircraft Production to CO, Felixstowe. *P.S.B. Armament.* 04.05.18. TNA AIR2/989.

37. Rennie, J.D. "Some notes on the Design, Construction and Operation of Flying Boats," Paper delivered to the RAeS on 18.01.1923. Paper published in the *Journal of the Royal Aeronautical Society.* 1923. P.123.

38. Request for Requisition to build Experimental Aircraft for the General Benefit of Aeronautics. Controller, Technical Dept. 02.12.1918. This Request contains the information re the proposed Atlantic Flight. TNA AIR 2/989.

39. Letter to Admiralty. *Transatlantic Flight*, 17.02.1919. Copy in TNA AIR 1/461/15/312/110.

40. "The Transatlantic Flight", *Flight,* 27.03.1919, P.402.

41. 'The Wreck of the "Felixstowe Fury"', *Aeronautics,* 18.12.1919. P.553.

42. Copy in TNA AIR 2/989.

43. *Flight*, 21.08.1919.

44. Letter to author 6 June 1989.

45. Rennie, J.D. "Some notes on the Design, Construction and Operation of Flying Boats," Paper delivered to the RAeS on 18.01.1923. Paper published in the *Journal of the Royal Aeronautical Society.* 1923. P.123.

46. *Ibid.*

47. See "Some Gosport Flying Boats for 1920", in *Flight,* 25.12.1919, P.1657. This article has sketches which show the proposed development of the Felixstowe boats into civilian machines.

48. Rennie, J.D. *Op Cit.* P.123.

49. *Ibid.*

Curtiss H-16 Bureau No.845 with post-war rudder modification. This machine was shipped to Pensacola on completion in October 1918. It was stored there until put back into service in 1921. Shipped to the NAF in 1922, presumably for storage, it was struck off in 1923.

Curtiss H-12B Convert N4339. Apparently converted on delivery in December 1917, this boat had the additional 'fighting top' armament. It was attacked by 10 German seaplanes on 15 February 1918, the boat having to flee the scene. Reported as worn out with a strained hull by July 1918.

Felixstowe F.2A N4304 assigned to Felixstowe.

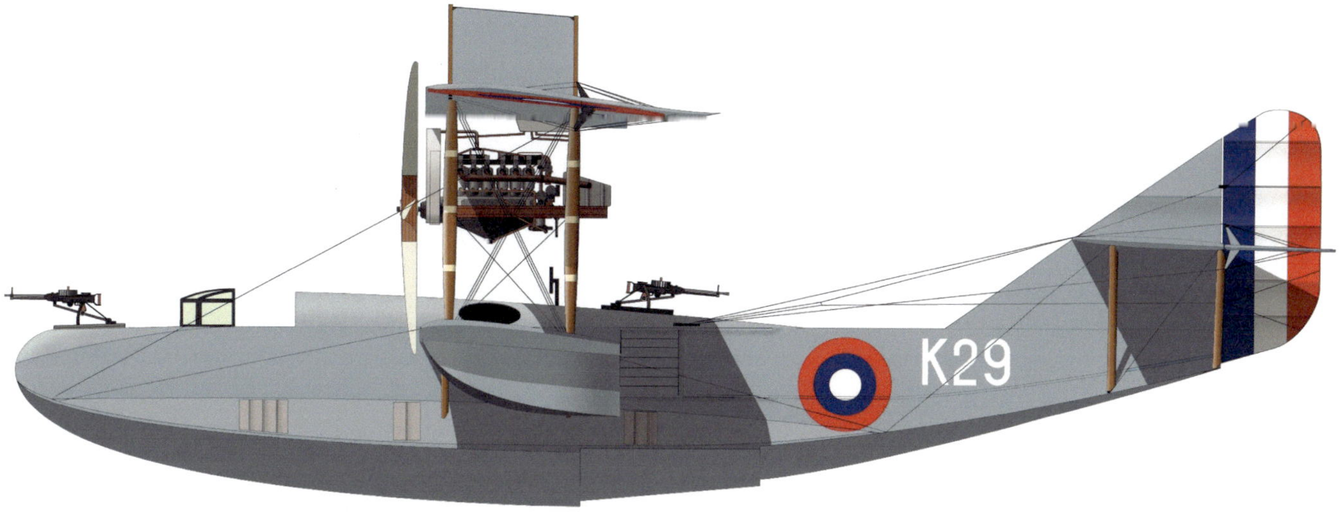

Curtiss H-16 K-29 was recorded at Killingholme on 16 September 1918, when it made a test flight. It had problems and the first patrol appears to be that of 26 October when it made a convoy escort of 440 nautical miles under Ensigns Kunkle and Rumill. It was still at Killingholme on 22 November when it made a Practice Flight under Ensign Kunkle.

8. Fighting Tops

Above: The cause of it all, the Germans introduction of Brandenburg two-seat fighting floatplanes to the North Sea. The first Brandenburg fighter was the W.12 and it set the standard for those that followed. The raised rear fuselage with the rudder below gave the gunner a superb field of fire. This machine, Marine Number 1183, bears the personal markings of Christensen. (AHT AL0064-13)

Below: The W.29 superceded the W.12 biplane and was the first in the line of successful Heinkel monoplane floatplanes that continued well into the post-war period. Marine Number 2536 bears one of the individual insignia that these aircraft from Zeebrugge and Borkum carried.

The US Navy was very interested in aeronautical developments that were taking place in the UK towards the end of the Great War as the UK was leading the world in the development of ship borne aircraft and traditional wood and fabric constructed flying boats. A report in July 1919 recorded that:

> *One of the F. boats was being used for experimenting with gun mountings, and a number of small cockpits for gunners had been placed in various positions in order to determine the best location for the gunner when firing over the tail. On the machine we saw the cockpit had been placed just to the rear above the left hand motor, the floor of the cockpit resting on the engine bearers which had been extended for this purpose. The position was found to be satisfactory, but, owing to the tremendous amount of pressure over the top plane, the work was made very difficult for the observer, and we were informed that the best position would be to place gun mounting cockpit on* (the) *trailing edge of the wing just to the rear of the central wing strut.*[1]

The gun positions referred to were to be called Fighting Tops, and were developed in response to the increased German Navy's aggressive actions in the North Sea on the introduction of Brandenburg seaplane fighter aircraft to the conflict. A 1919 Report on the anti-submarine war states that German seaplanes were observed to be able to alight on the open sea to await the *Large America* patrols with the certainty of getting off again. They completely eclipsed the *Large Americas* in power of manoeuvre and fighting qualities, and they possessed "reliability" and "considerable endurance."[2]

As early as June 1917, the ascendency of the new German seaplanes was being discussed. It was reported that the Germans had lately produced seaplanes that had considerably better performance than the British. These machines were believed to be lightly constructed and not up to the British standards of strength and "it was no doubt for this reason that they got their extremely good performance, but the fact remained that with these seaplanes they were becoming more and more bold in their reconnaissance patrols, attacks on shipping, etc., and our own seaplanes seemed powerless to prevent this." The German floatplanes had better performance than any of the seaplanes now at Dunkirk. The result being that the British low height submarine patrols had practically stopped, although aeroplane patrols operating at great height above the coast had not been interfered with. It was strongly recommended that *Large Americas* be supplied for this purpose – a fighting escort of Sopwith Pup landplanes being provided from Dunkirk. Facilities for operating *Large Americas* from Dover or Dunkirk were not good. If the scheme using *Large America* boats was approved the escorting Pups would require an air bag in the tail. The question – Should a small fighting seaplane for fighting purposes be developed? – was raised. Also as there were not enough *Large Americas* for this work it was suggested that

Above: A Brandenburg W.29 attacking downed *Large America* flying boat N4305 of Capt E.A. Mossop. On 31 July 1918, the boat was attacked by five W.29 monoplanes of 1.C Staffel and forced to alight. Shot up on the water it caught fire and sank. Three of the crew were rescued but two drowned. The action was captured by a cameraman in one of the W29 floatplanes. See Chapter 3 for a full account of this action. (AHT AL0588-014)

Above: The Curtiss H-12 initially had spigot mounts for the front gunner's Lewis gun. Other guns were mounted where possible but protection from the rear was lacking. Note the airscrews rotate in different directions.

land machines could be used.[3] This discussion was to continue into 1918 and eventually led to the design of the N.1b fighter flying boats.

The Admiralty wrote to the Vice Admiral, Dover Patrol, in late February 1918 that their "Lordships have had under their consideration attacks recently made by hostile seaplanes upon Large America Seaplanes engaged in carrying out the North Hinder anti-submarine patrol from Felixstowe." The Admiralty wanted fighting aeroplanes from Dunkirk to arrange to meet the Felixstowe patrols and act as escort. Despite having raised the question of Dunkirk and Felixstowe co-operating before, "up to the present no patrols from Dunkirk have been provided on the ground of no machines being available." The senior Naval Officer, Dunkirk, "should now be directed to hold a fighting squadron of D.H.4s in readiness for this service."[4]

The Senior Officer, Dunkirk, Capt C. Lambe, wrote in reply that according to his instructions from the Admiralty the Air Ministry laid down the composition of the Air Wing at Dunkerque *lent* for serving with the Navy was to be composed as follows *(Author's italics)*:

- One squadron of D.H.4s for Fleet Reconnaissance and Photography.
- One squadron of D.H.4s for anti-submarine patrol work.
- Three squadrons of fighting machines.

Lambe then noted that he assumed that the Admiralty was referring to the latter D.H.4 squadron *that was being formed.* While acknowledging that the D.H.4 would be able to co-operate with *Large America* boats from Felixstowe, he pointed out that "the D.H.4 type of machine is not a very suitable type for attacking hostile machines although extremely suitable for defensive measures." Also, it would be necessary to ascertain the "exact limits, times, of starting and leaving, and duration of the patrols carried out by *Large America* Boats operating from Felixstowe, but even then the chances of the two machines leaving Dunkirque and Felixstowe respectively ever sighting one another at sea are extremely remote." He noted that the D.H.4 anti-submarine patrols from Dunkirk would be able to carry out a greater number of patrols then the *Large Americas* from Felixstowe and that this "would appear to be worthy of consideration."[5]

Lambe came under heavy criticism in March for his "non-

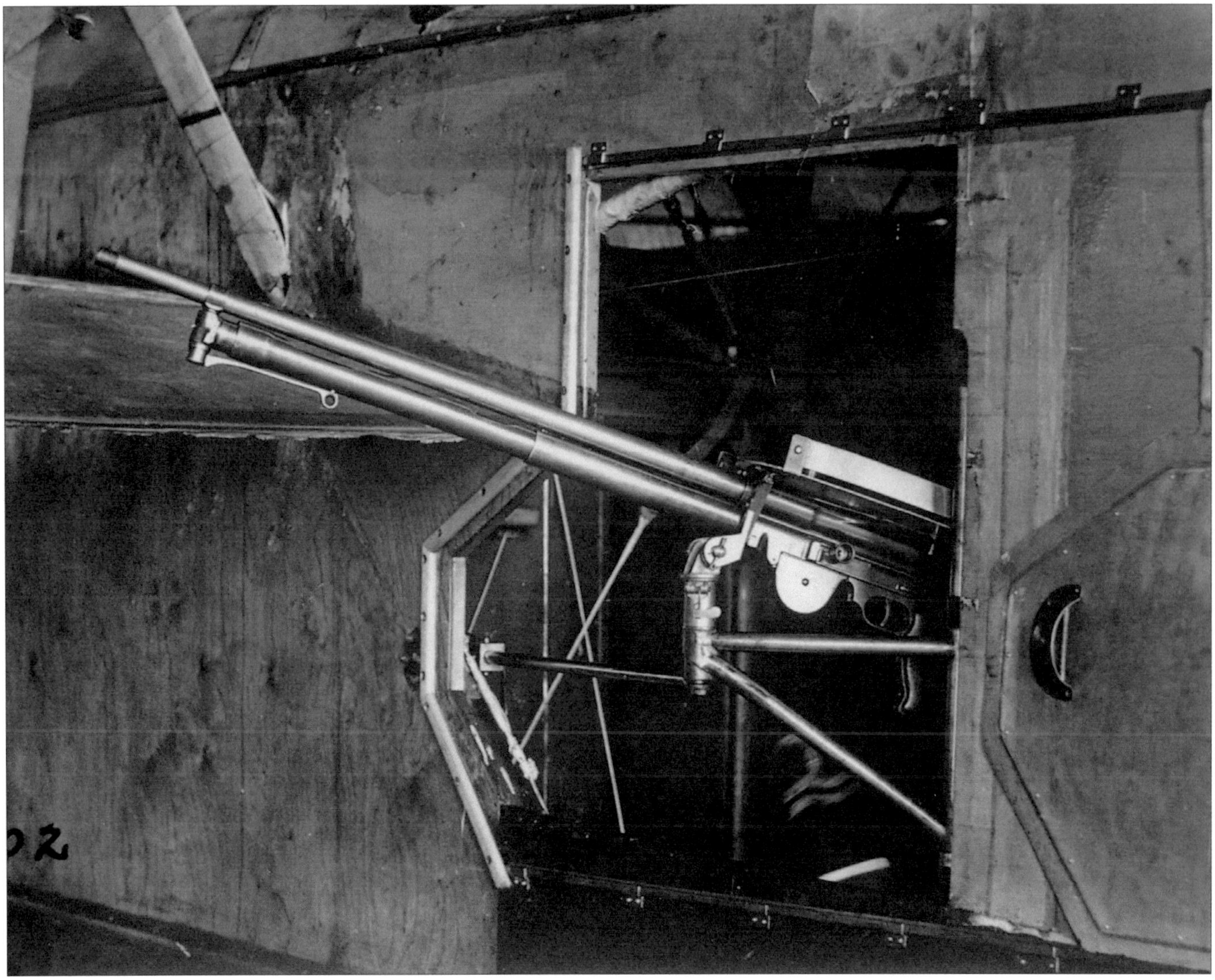

Above: The standard Lewis gun on a Gallows mount as fitted to the F.2A, H-12 Convert, H-16, F.3 and F.5.

compliance with Admiralty directions" and a letter concerning this was drafted relieving him of the command of the Naval Air Service units in the Dover-Dunkirk area. Apparently not sent as a hand written note in the file states that Admiral Keyes decided the letter should not be sent and "in any case, existing conditions are now different from when the proposed letter was drafted."[6] To date nothing has been found to confirm that any co-operative flight between Felixstowe's boats and Dunkirk's D.H.4 biplanes ever took place.

In March 1918, John Porte wrote that "our seaplanes have been considerably worried lately by attacks from enemy aircraft." Steps had been taken to combat this, but had to wait until sufficient flying boats were available. The patrols were sent out in flights of three when on extended patrols, and when hostile aircraft were sighted they were to take up a V formation with the leader being at the apex. "Machines are being constantly exercised in this formation, and considerable improvement has been obtained." Porte was against fighting land machines flying in convoy with the flying boats to protect them from attack. He considered that in order to operate successfully, a *Large America* seaplane "should be able, and I am satisfied can, take care of itself."

Porte was worried that "once the idea gets loose amongst the pilots, etc., that these machines are very vulnerable to attack, a rot sets in, with the result that the machines run away as soon as enemy aircraft are sighted. However, it is only necessary to prove that these machines could take care of themselves very successfully, for the whole point of view to change in this matter." Porte had sent out a flight under Flt Commander Galpin with orders to engage the enemy and this he had done, "fighting five enemy seaplanes with only two of our own, and by winning the action, proving beyond all doubt that the Large America Seaplane, when properly handled and flying in formation, are well able to defend themselves against interference from enemy aircraft." The result of this action "has been to completely change the opinions of the personnel

Above & Facing Page: Lewis guns on Gallows mounts stowed inside a Large America.

generally."

Porte proposed to send three machines fitted with "special armament" on days when, from experience and reports, it was fairly possible that enemy aircraft would be encountered. This special armament comprised

> *two Lewis Guns mounted in the forward mounting, one on top of the cabin for the Stand-by pilot, three in each of the Gallows mountings on each side aft, and two on the aft central mounting, making a total of 11 Lewis Guns. A crew of six will be carried on these occasions.*

Porte was convinced that three boats with this armament should be able "to tackle enemy aircraft up to twelve in number with every chance of complete success." He also reported that he had requested a 6-lb Davis Gun for use against aircraft.[7]

The question of larger armament for the *Large America* seaplanes was being discussed almost as soon as they were received. The question raised at a Committee meeting in May 1917 was whether a larger gun than 0.303 was desirable. It was stated that the machine gun was needed for rapidity of fire. A large bore gun would be desirable if "equal rapidity of fire could be guaranteed." One hundred 1½ pdr Guns were on order but were for rigid airships only. They were belt fed and would take 50 rounds per gun. It was agreed that experimentation should be undertaken as the German seaplanes might mount larger guns "and we should be prepared to meet this." It was generally agreed that concentration of fire from the maximum number of guns was best.[8] The Germans did mount a 20 mm Becker cannon on Brandenburg W.19 *Marine Nummer* 2237 as an experiment.[9] Although another 23 W.19 aircraft were ordered with this armament, as far as is known this was not used operationally.

While the Felixstowe boats did encounter the Germans and some successful engagements took place, Porte's report did not mention the fundamental problem with the *Large America* flying boats, their vulnerability to attack from below. While he had improved this by canting up the rear of the hull and installing the Gallows mountings[10] at the sides of the boat, the F.2A was still vulnerable to an attack from behind. In any engagement the boats had to fly low to the water to prevent the enemy from being able to get into the dead spot behind and below. The defects of the *Large America* boats as fighting machines were detailed in a later report.

> *The present arrangement of gun positions consists of a gun mounting in the bow which commands almost all the forward hemisphere and a fair part of the rear, and three gun mountings in the fuselage* (sic) *behind the planes, which together have not*

a suitable field of fire to protect the machine from an attack from the rear.

This arrangement requires a crew of five, but the gun positions are so placed that the advantage of superior numbers is lost and fire control is bad.[11]

The problem with a *Large America* boat acting alone had been demonstrated on 31 July 1918. F.2A N4305 from Yarmouth was on patrol in the neighbourhood of Smiths Knoll when it was attacked by five enemy Brandenburg W.29 monoplane seaplanes in formation. Two on the starboard bow opened fire as they closed in and the W/T operator, Pte Ronald R.N. Cooper was hit almost immediately and killed. The three other enemy machines that were sighted on the port bow attacked from a position on the tail of the boat. Within a few seconds of the attack the F.2A had been hit in the gravity tank and forced to land. Capt E.A. Mossop, the first pilot, ordered the crew out onto the wings and the enemy attacked again in line ahead from the rear, the F.2A burst into flames. The engineer, 19 year old AMIII William H.D. Dingley disappeared and it was presumed he fell back into the flames and was killed. Capt Mossop, Lt G. Hodgson and the Gunlayer, Pte Greenwood, were all in the water and the latter, whose lifebelt refused to inflate and who could not swim, was supported by the two officers for 35 minutes until they were picked up by HMS *Halycon.* When the boat sank it left a large patch of burning fuel on the surface that the survivors had a great difficulty in keeping clear. The enemy machines had a speed a good deal superior to that of the F.2A.[12]

In the period from 1 April 1917 to 26 August 1918, in 19 fights, nine German machines were destroyed and 15 of his personnel killed, for the loss of eight British boats with 14 killed, ten interned and three POWs. "In spite of the bad gunning of the boats, their relatively slow speed, the lack of trained gunners, and superior numbers of enemy aircraft, "the record is reasonably good." In addition, the boats in this area bombed a number of submarines, destroyed two Zeppelins, and carried out useful reconnaissance.

In some degree the lack of success in attacks made by the enemy is due to the inexperience of his pilots, but this factor in our favour cannot be expected to continue.

The most important factor was considered to be "concentration of fire."

It is not sufficient that the blind spots be small, the gunners must be able to support each other...The big machine has the added disadvantage of being a large target, and owing to its slow speed, it is unable to choose the time and method of attack.

It was considered that for a successful defence of the boat or boats in formation:

Above: This gunner demonstrates the position adopted when under attack. The Felixstowe boats had rear gun positions that the Curtiss H-12 lacked but even so it was still vulnerable from below and behind.

Facing Page: Front gunner's cockpit on a Phoenix-constructed F.3 hull with Scarff gun ring mounted. The gun ring has a yoke for twin Lewis guns. Note the bomb sight, Phoenix logo and the gloss finish of these boats.

1. At least two gunners should be able to bring their guns, preferably twin guns, to bear on an attacking enemy.
2. This must be achieved without manoeuvring or breaking formation.
3. There should be continuity of fields of fire.
4. Good fire control was essential and the gunner must be in communication with the pilot.
5. Good intercommunication between boats when in formation.

The F.2A now in service was "overloaded and additional gunlayers or armament can only be carried by cutting down the range of flight." The low speed was not considered so important as the boats could not overtake or get away from light enemy fighters. It was hoped that the F.5 that was entering production would have the performance of the first experimental machine as it was "not affected to the same degree by the question of weight and speed as the F.2A." The great difficulty in arming flying boats was that the upper plane was so high above the top of the hull that gunners in the bow and rear of the hull were separated and fire could not be concentrated nor controlled. The Gallows mountings on the sides of the Felixstowe boats were considered inefficient as they were difficult to shoot from and could only carry one gun, and could not be seen by the pilot or front gunner.[13]

Porte was also criticised by the Director of the Air Division who, while concurring that the *Large Americas* should be self-protecting, noted that

> *The Commanding Officer of Felixstowe Air station has not apparently quite grasped the idea of sending out fighting machines from Dunkirk which was in the first place with an idea of destroying the enemy and to avoid Large Americas being risked and having to fight for their information. There is no question that in the formation of three they can defend themselves, but in order to gain the necessary control and speed for fighting they are forced to drop their bombs, and have at least on one occasion missed an opportunity thereby of attacking an enemy submarine.*[14]

As mentioned before, some officers at the Admiralty had no practical knowledge of the circumstances of aerial warfare at sea and the problems of arranging for scouts to meet up with flying boats at a precise location over the ocean where there were no landmarks. A note to this Minuet stated that "No permanent alterations to large Americas should be undertaken which would impair their special capability to make long reconnaissance flights." This note shows how important these flights were and that the Admiralty would go to great lengths to

AIRCRAFT DEPT
BRADFORD
QUALITY
FIRST

Above: 8661 H-12 Convert with a F.2A hull. The side windows aft of the wing are the indicator of the change in hull. This machine was delivered to Felixstowe in January 1917. It had an eventful life being involved in submarine attacks on 23 April, attacks by enemy seaplanes in December, and was up for deletion by the end of the month. Given a new F.2A hull the conversion was completed by 6 January 1918. 8661 was involved in two fights with enemy seaplanes on 5 February and 12 March and the crew were credited with an EA shot down on each occasion. In October it was reported as worn out but was not deleted until the following January. This photograph accompanied a report on "Fighting Tops."

continue them.

The specifications for the Admiralty N.3 and N.4 flying boats had been issued and these were in development. These were to eventuate as the Short Cromarty and Vickers Valentia for the N.3 Specification, and the Fairey Atalanta and Titania to the N.4 Specification. Although work continued after the Armistice they remained prototypes only. It was still the opinion of some that a fighting seaplane should be developed. and this had been discussed at a Special Meeting of the Technical Committee of the Admiralty Air Department on 30 May 1918, the topic being the "Consideration of the desirability or otherwise of designing a Fighting Boat Seaplane for Escort Duties with Large Flying Boats." The reason for the Meeting was presented by Major D. Hallam, representing Felixstowe Station. He stated that the "Germans had developed a fairly fast single-seater seaplane scout which was apparently about 5 knots faster than the F.2A Boat. These were flown in groups of five, and usually lay on the water on a calm day on our fairly well-known routes and started up when they heard our engines. This had cut down the efficiency of our patrol considerably. Instead of being able to send out single machines, we had to send out 3, and it was quite probable that in the near future we should be obliged to send out 5" and that this group should be met by a special escort. Hallam noted that up to the present not much attention had been paid to the "Fighting Flying Boat."

Lt Col A. Ogilvie stated that if a landplane twin-engined bomber could do the work, "there was nothing further to be said." In addition to the fighting seaplanes there were fighting aeroplanes that went up from the German coast and "it frequently happens now that a reconnaissance was not carried out because the Boats found themselves up against a superior German formation." If it was possible to send out three boats, then they could "win through" but it was considered that by the end of summer it would be necessary to send out five boats, therefore it was necessary to provide some sort of escort.

It was considered whether the Felixstowe Boat could have its armament strengthened "to make the machine capable

Above: Front view of Curtiss H12B N4339 with two nacelles at the inner bay interplane struts. The nacelles extend above the upper surface of the top wing. All gun rings are armed with twin Lewis guns. Note the dazzle painted F-boat in the background. N4339 was part of an order for 24 flying boats from Curtiss in the USA. Received at Felixstowe in December 1917, and was given an F.2A hull to become an H12 Convert. On 15 February it was attacked by up to ten enemy seaplanes and was forced to flee the engagement, fortunately without any injury to the crew. By July it was reported to be worn out but was still at Felixstowe in January 1919.

of pushing its way through." The De Havilland D.H.10 was considered "as good a fighting machine as the D.H.9 although not so manoeuvrable." The twin-engined Caudron that undertook mock-fights with a Sopwith Camel, a Mono Pup, and a Bristol Fighter "gave an extraordinarily good show, and it was found that, due to the pilot being near the rear observer, it was possible to keep the scouts off the tail." This Caudron would have been the R.11, two of which had been obtained from the French for evaluation.[15]

Capt C. Dalison stated that he had a good deal of experience in flying land machines over sea, and considered that as long as there "was someone who could pick the pilot up in the event of engine failure, that would form a very strong argument for concentrating on escort of land machines, the escort having his safety looked after by the boat he was escorting, in the case of engine failure,." In reply Major Hallam pointed out that land machine pilots would not be able to make the long patrols of 300 miles or more. "There might be fog and they might lose each other on numerous occasions. At present the boats came back singly on practically every occasion."

After further discussion it was decided that these was a clear case for a fighting boat seaplane. Major Hallam raised the question of formation flying and stated that in the past there had not been enough boats available to enable practice to be carried out. He was of the opinion that the fighting capacity of the present boats could be greatly improved.

In conclusion the Chairman stated that there had been a good deal of difference of opinion at the different stations. "The whole question of equipment should be looked into and the combined experience of all stations put together and a satisfactory scheme arranged."[16]

It was not just the long-range reconnaissance flights that were in trouble. In July Col Samson, Commanding No.4 Group, SE Area, wrote that "in view of the continued interference with our Anti-Submarine Seaplane Patrols from Felixstowe, at least two Flights of very fast fighting type seaplanes are urgently required at Felixstowe to accompany the Anti-Submarine F.2.A and F.5 seaplanes."

Above: The nacelles on the F.5 (N90?) were carried completely above the wing. Compare with those on the N4339. Although not indentified in the reports, this F.5 is sitting in the beaching trolley for N90.

"At present the situation is – that our anti-submarine patrols have to be sent out in formations of 5 to 6 F.2.A's as the hostile aircraft always fly in a formation of 7. The Hostile Seaplanes are very much faster and handier than the F.2.A's." It had been possible to accompany the patrols with aeroplanes "but this is not the correct method and suffers from many disadvantages."

"The Fighter Seaplane should be at least 100 knots with 3 hours fuel at full speed at surface level. This gives them nearly four hours fuel at cruising speed. The method of carrying out patrols would then be 2 F.2A's with bombs and 3 to 4 Fighters." It was requested that a floatplane fighter with a Rolls Royce Eagle VIII would easily surpass the speed of the Germans "who only use engines of approximately 200–260 H.P."

The letter concludes with the observation that the "very large machines of slow manoeuvring power cannot successfully be fought against smaller, faster and handier machines."[17] Admiral Hickley agreed with this assessment and added that the *Large America* flying boats were at a disadvantage against the fast enemy machines that could choose to run or fight according to the size of the British formation. The provision of two or three flights of such seaplanes as proposed by Samson "would give us a fighting supremacy over hostile machines on all occasions, which is most desirable."[18]

The problem had been discussed at a Meeting on 17 July. Cdr Longmore stating that there was a requirement for a small twin engined flying boat with a high performance for work in the southern part of the North Sea and parts of the Mediterranean "where a comparatively short fuel range was sufficient but where hostile aircraft might be encounters. There was at the moment a distinct shortage of a high performance type of submarine reconnaissance seaplane capable of tackling German seaplanes in these localities." After discussion it was agreed that the machine was required and that a specification be prepared. Cdr Longmore then proposed a further new specification for a twin engined fighting machine along the lines of the F.2 or F.3. This machine was to have guns of larger calibre than .303 and was to be used as an escort for formations of *Large America* boats operating in the North Sea. The Coventry Ordnance Automatic Gun appeared to be the most promising and he proposed that the machine be ready for installation of the gun as soon as trials were completed. Comdr Porte considered that the F.5 "which had just been laid down at Felixstowe could very well be adapted to take two of these guns. This machine would have a better performance than (the) F.3

Above: Frame work of the F.5 nacelles used on the F.5. The wing structure did not have to be altered to fit these nacelles.

and should be ready in about two months." The proposition of using the guns for attacks against submarines as well as enemy aircraft was considered but the decision was that the machine should be used purely as an escort fighter. The performance of this machine should be better than the performance of the bombers it was intended to protect. It was decided that a new specification be drawn up and that it should replace the N.3.a specification for torpedo seaplanes.[19]

The unpleasant truth that the Curtiss and Felixstowe flying boats were no longer masters of the North Sea was reached by the end of July when the Admiralty reported that due to recent actions it was necessary to employ not less than six flying boats in order to undertake a reconnaissance of the southern part of the Heligoland Bight. The flying boats were slow to manoeuvre and very vulnerable to attack by enemy seaplanes. It was suggested that the following methods could be adopted:

1. Reconnaissance with a less number of boats supported by aeroplanes.
2. Reconnaissance with a less number of boats escorted by fighting seaplanes.
3. Reconnaissance with fighter-reconnaissance-seaplanes. These being capable of self-defence, should probably be employed in pairs.

"It would seem desirable to develop two types, a fighter seaplane and a fighter-reconnaissance-seaplane." The letter ends by pointing out that the Germans had successfully produced both types of these seaplanes.[20]

In the meantime the experiments to improve the defensive power of the flying boats were carried on. Felixstowe was to build the nacelles on the wings and Grain would build the nacelles at the back of the engine mounts. Felixstowe also modified one boat to have a single nacelle in the centre section of the top plane. It was to be the end of August before the Air Ministry replied that modifications

are being tried which it is considered will render the Flying Boats much more capable of self-defence.

In addition, and as an immediate step, arrangements are being made to fit six Fairey Type 3b Float Seaplanes with Rolls Royce engines instead of the Maori Sunbeam type. It is proposed to test these on service in the North Sea. It is feared that their fuel range will prevent them from operation to the limits at present covered by the Boat type Seaplanes, but their speed and increased handiness may give them sufficient fighting value to meet German Seaplanes at intermediate distances.

It was proposed to reduce the number of flying boats supplied to Felixstowe and Yarmouth by three on the supplying of the Fairey machines.

In addition it was proposed to convert three De Havilland D.H.9 biplanes into floatplanes, and three other D.H.9 machines with "emergency flotation gear", for the purpose of comparative trials.

It was acknowledged that the development of a Fighter Seaplane and a Fighter Reconnaissance Seaplane type was

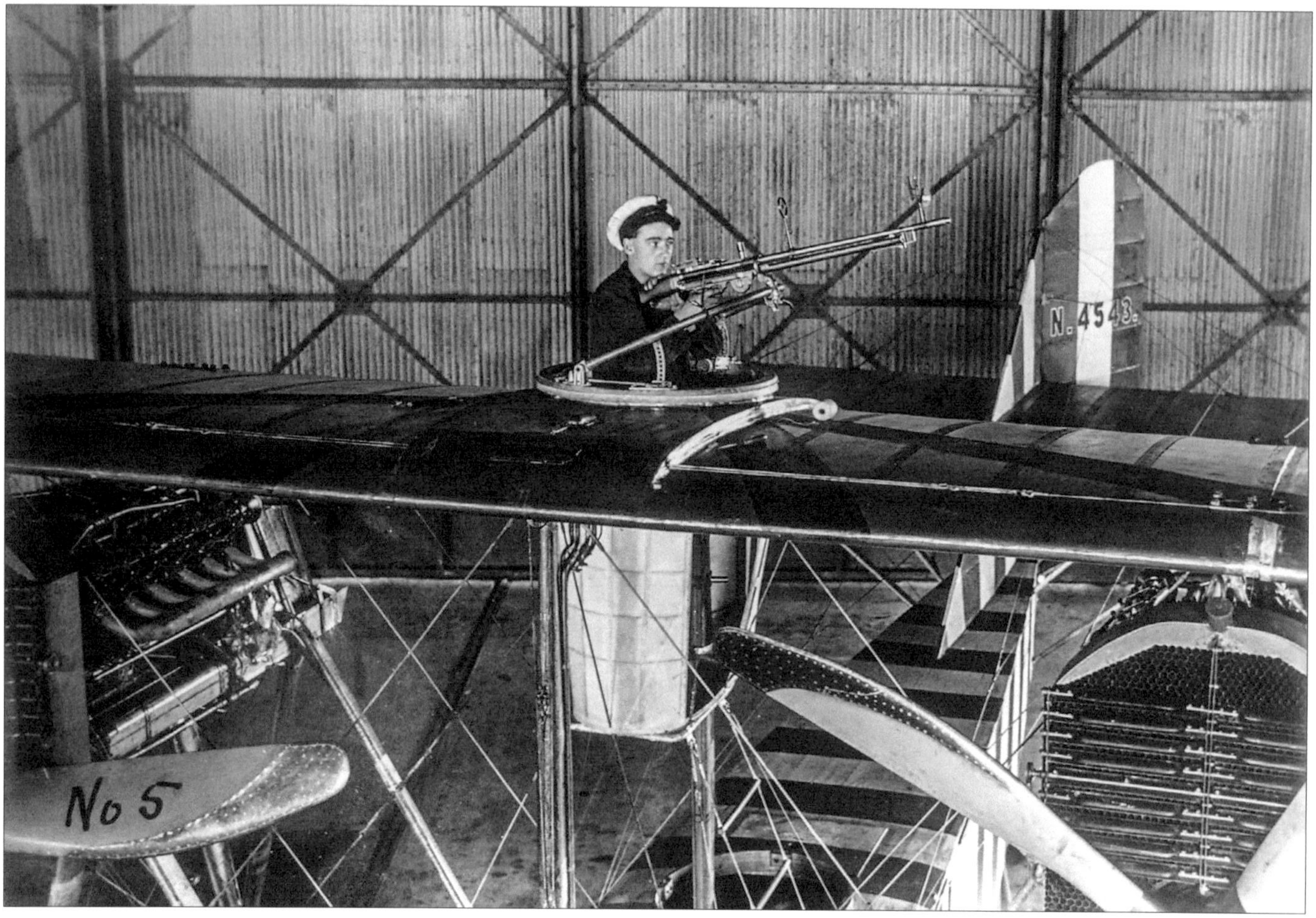

Above & Facing Page: Demonstrating the Fighting Top on dazzle painted N4543. This nacelle did not allow the gunner to slide the gun ring forward and backwards. Note the open side and rope ladder access. N4543 was with No. 230 Squadron, RAF, from July 1918. It was lost on 9 November when forced to alight due to running out of fuel. The seaplane sank but the crew were picked up by Dutch trawlers.

required and specifications would be drawn up as time allowed. The N.4 specification for the Large Protected Flying Boat Seaplane had been let and the type was now under construction.[21] In addition Porte had the triplane flying boat under way that was to appear as the Felixstowe Fury.

The proposals were discussed at the Air Ministry on Saturday 7 September 1918, and the Technical Department considered that the "design of such a machine to operate at the distance into the Heligoland Bight which are required, presents considerable difficulties. It is not certain if the design is practicable." As noted on the undated report "Armament of F.2.A Flying Boats" reproduced below, the concept of a fighting flying boat was thought unlikely to be successful.

Top plane gunner nacelles, or "Fighting Tops", were developed to enable the *Large Americas* to defend themselves without a great loss of performance. It was immediately evident that the idea could be applied to other large aircraft, particularly large bombers. The idea that the aircraft of World War I were fragile and not designed in a scientific way has become accepted fact to the general public. In reality even the venerable B.E.2 was designed in accordance with the science of the day and was proof load tested. Similarly, the development of the Fighting Tops was not a haphazard arrangement but subject to trials to determine the best method of providing protection. A wealth of reports on the trials has survived and they give a good example of how the problem was tackled and overcome with a mixture of design and practical application, and therefore the reports are reproduced in detail.

The design and fitting of the top-plane nacelles had been carried out by the Experimental Department at Felixstowe. "The two oval-shaped nacelles are built round the top of the interplane struts, next to the engines, with the ends almost flush with the leading and trailing edges of the plane; they are about 30 inches deep and 2 feet wide, the top coming level with the upper surface of the plane. On top of the nacelle is a square frame, carrying a Scarff Ring Gun Mount; this frame can slide along the length of the nacelles, thereby allowing the gunner to actually fire vertically downwards over both front and rear of the planes."[22]

A mock fight was conducted by a H.12 fitted with the top

plane nacelles and two scouts, a Sopwith Camel and Dolphin, and two two-seaters, a Bristol Fighter and a Sopwith Bulldog. The report on this experiment reads as follows:

The attacking machines had arranged to make a concerted attack on the Flying Boat, the general scheme being for the Bristol Fighter and Bulldog to occupy the Flying Boats attention while the two Scouts dived on it from above: unfortunately their initial effort was wasted on another Flying Boat which went up the coast just previously. When the mistake was realised matters were further complicated by the Dolphin's engine overheating and the Flying Boat keeping two far out to sea to give the land machines a fair chance at such a low altitude.

Above: This photograph is recorded as showing the position on F.3 N4248. The gunner's position is very rudimentary with just the small platform to stand on. Access was by the rope ladder. Nothing further is known about this particular attempt.

Facing Page: Detail of the nacelle on F.3 N4409 mounted at the wing trailing edge on extended engine supports.

GENERAL IMPRESSION.

It was quite evident that the attackers could have made much more use of their superior performance and numbers under more favourable conditions, but they put up quite a good enough show for the top gunners on the Flying boat to study its armament fairly completely.

Both were very enthusiastic in their approval of the scheme of armament which in relation to that previously carried on a boat of this type has certainly altered it from a fairly simple proposition for one to two well managed light E.A. into a very serious problem for twice the number to attack. As far as can be judged from this flight, all that was hoped from this scheme of defence would appear to have been most fully realised.

FIELD OF FIRE AND VIEW

As far as the top plane gunners are concerned the boat is almost ideal in this respect as they can shoot vertically downwards over both the leading and trailing edge of the planes, so that there practically is no area which is blind to both of them. In addition the forward gunner commands almost all the forward hemisphere and a fair part of the rear; a rather important shot

No7

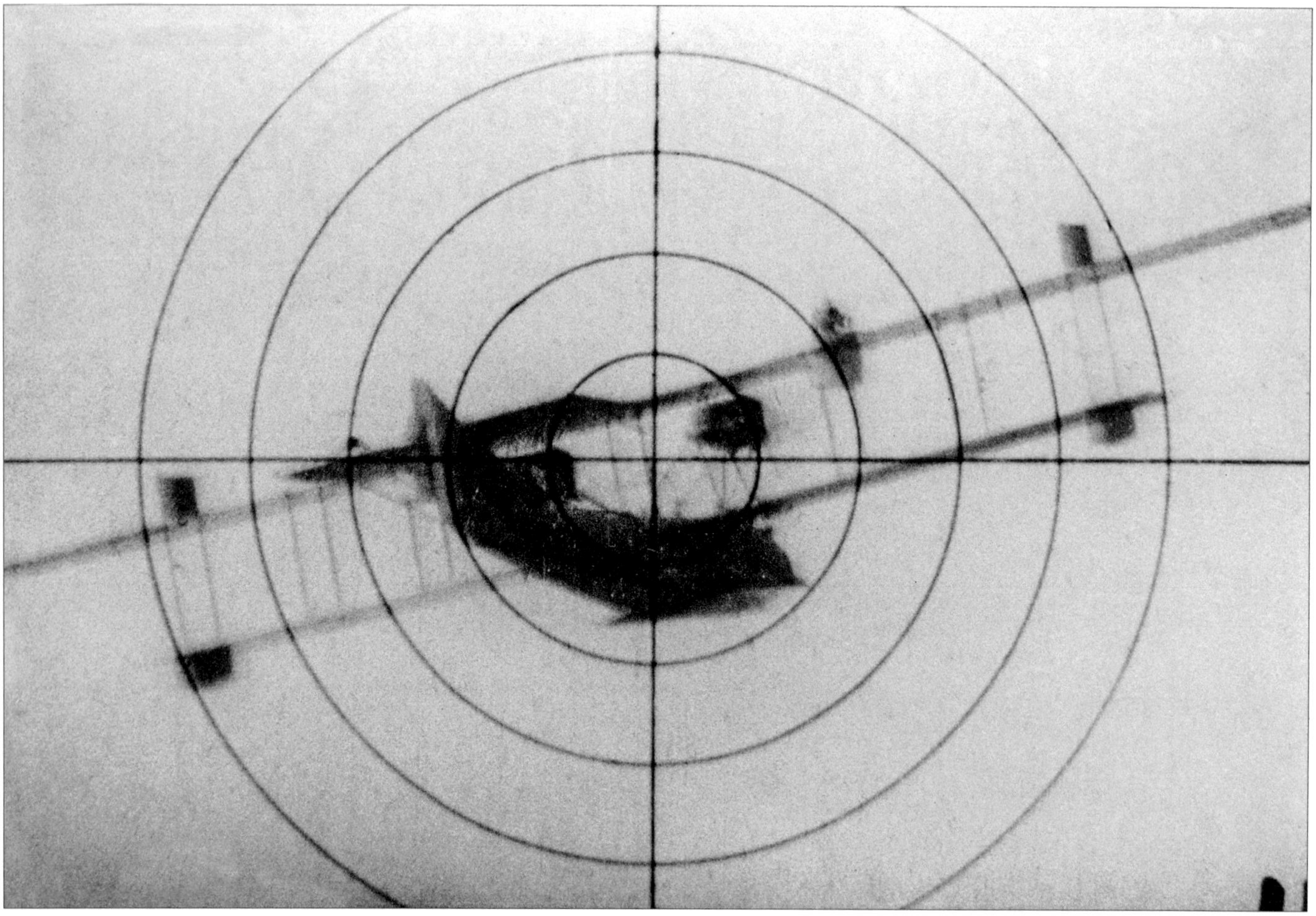

Above & Opposite: Camera gun photographs showing N4339 and how the attacking aircraft is in the gunner's line of fire and vice versa.

for him under the planes is interfered with by drift wires from the bows.
In the fight the boat was attacked from every quarter though never simultaneously be more than two machines at once. The remarkable absence of blind spots was most obvious to both top plane gunners and quite beyond any of their previous experience in fuselage cockpits of other single or twin engined machines. The observers in both the Bristol and Bulldog each took a film with a camera gun, in nearly all of which the two gunners on top of the Flying Boat could be seen with an unobstructed aim. The pilot of the Bulldog had a cinema camera which unfortunately only functioned for two exposures, both of which were taken under the tail of the H12 in full view of both the gunners.

It was further observed by both the gunners that they would be more likely to hit an enemy approaching from behind the tail if the pilot flew straight, and that any form of manoeuvre would be more likely to hinder them than afford any assistance. In action, therefore it would seem advisable for the pilot not to bank, side-slip, etc, any more than he feels necessary to avoid being hit by the enemy.

It is worth noting that should the machine armed in this manner encounter a large enemy machine of inferior performance, the latter would have little hope of surviving the combined forward fire of the three gunners however it tried to escape.[23]

The best results were obtained from the two nacelles in the top plane. The objection to the two nacelles was the loss of performance which was "certainly considerable" but it had "not been accurately measured." It was suggested that the Felixstowe modified boat go to the Isle of Grain for performance tests. "At the same time in view of the very good field of fire which is obtained by this alteration, and also of the fact that control of the boat is not noticeably worse I think the alteration should be worth while doing as an experiment only on a land machine as it might lead to valuable data being obtained on the design of future big machines." It was suggested that a D.H.10 with only the pilot in the fuselage be tried. "There may be objections from the structural point of view" but it was thought worth the experiment.[24]

Another series of trials were carried out in September with a Maj Jones, Lt Banister, and Maj W.C. Mitchell, Chief Experimental Officer, of the Orfordness Armament

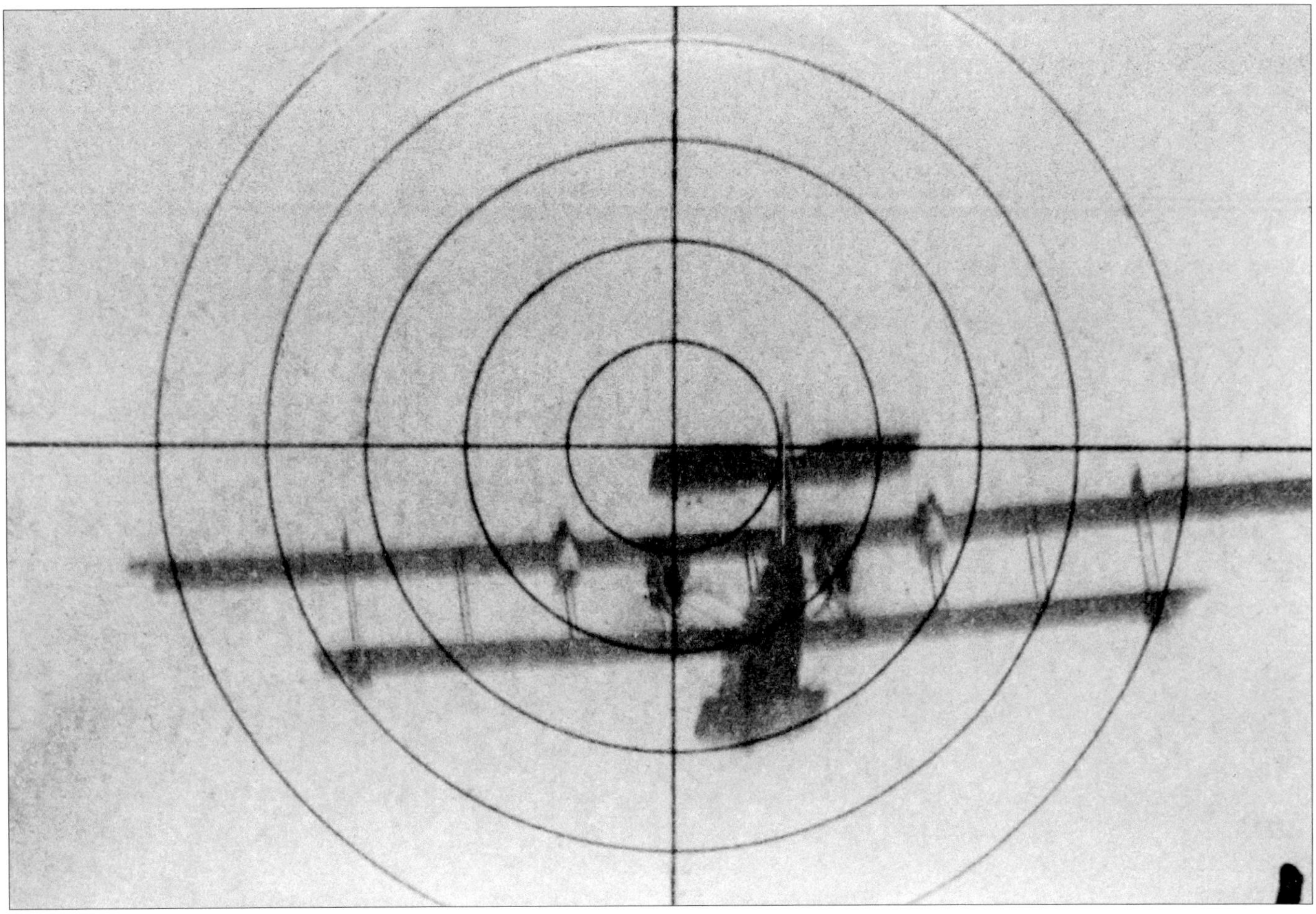

Experimental Station.[25] The following three arrangements of Fighting Tops were tried:

(a) F.2A No.N4339. Two gunner's nacelles let into the top plane immediately over the pair of interplane struts outside the engine. A Scarff ring is mounted in these nacelles, and is made to slide fore and aft along the length of the nacelle. This enables the gunner to fire either over the leading edge or over the trailing edge.[26]

(b) F.3 No. N4409. Two nacelles built on the back of the engine nacelle. These nacelles are provided with a Scarff ring level with the top of the plane, but there is no movement of the Scarff ring fore and aft.

(c) F.2A No. N4543. One gunner's nacelle let into the top plane at the middle of the centre section. This nacelle is provided with the Scarff ring, but there is no fore and aft movement.

(a) & (c) were flown to Orfordness and engaged in a Mock Fight with the following land machines: Bristol Fighter, Bristol Arab Scout, Rolls Martinsyde Scout and two Camels. Owing to engine trouble (b) was not able to proceed to Orfordness.

A concerted attack by the 5 land machines was put up against each of (a) & (c) in turn, and the following points were brought out:

F.2A No.4339 with two top plane nacelles. These nacelles were considered very satisfactory. At no period did the gunners lose any machine in a blind spot, and the ease with which all machines could be kept under observation was particularly noticeable. The intercommunication by means of Klaxon Horns for calling attention was very satisfactory. The nacelles should be deepened about 3 or 4 inches, and a quickly operated release should be provided for locking the fore and aft movement of the nacelles.

The ease of manipulation of the mounting is very much increased by the shock absorber which balances the fore and aft travel.

F.2.A. No.4543 with single nacelle in top centre section. This was found very much inferior to the other top plane mountings for the following reasons:

(1) The field of view is very much restricted downward as there is no fore and aft movement possible. At several periods during the fight machines were lost to view for a considerable period under the wings and one could not tell whether they were coming out in front or behind.

(2) The field of fire is also further restricted by the propeller tips which project above the top plane, and

Above: Detail of one of the outer nacelles. There is no indication of how the gunner made his entry to the nacelle.

then shooting backwards, by the wires and bracings of the tail.
(3) One appears to get the full blast of the slipstream to this nacelle which makes the manipulation of guns very much more difficult than in the other top plane mountings.
(4) The intercommunication between the gunner and the pilot was bad. The gunner can only see the front observer, and he had to transmit any signals to the pilot.

F.3 No.4409 with two gun nacelles behind the engine. This machine was not used in the Mock Fight as it had engine trouble in the afternoon. Flights were taken later in the machine. This arrangement had practically all the disadvantages as mentioned in F.2A. 4543 above, the effect of the slipstream being even worse. The field of fire forward was still more restricted. Although only one nacelle was fitted, it was assumed that this scheme entailed two nacelles, one behind each engine.

Under "General Conclusions" the report stated that it was *of the utmost importance in all questions of defence of machines that the maximum number of guns shall be capable of being brought to bear simultaneously over the largest possible area. All questions of fire superiority depend upon this concentration of fire.*

For this reason the two top plane mountings were considered far superior to either of the others. The ability of the boat to defend itself was "at least three times as great as that of the ordinary boat, and when compared with (b) or (c) was considered more than twice as good.

In addition to the question of concentration of fire, the enormous superiority of type (a) from the point of view of lookout and the ability of keeping all the attacking machines under observation was "very definitely brought out in these trials." The gun arrangement was considered superior to the others and the intercommunication between the three gunners and the pilot was very good in type (a) and definitely bad in the others.

It was acknowledged that there would be a loss of performance with the nacelles fitted, and the conclusion was that if any type of nacelle was to be fitted, then type (a) was

Above & Below: The nacelles on this boat are above and below the wing like those on N4339. The number applied to the nose was usually the last two numbers of the machine's serial. "15" does not fit with any known serials of boats fitted with Fighting Tops. Unlike N4339 this boat does not have a cockpit canopy.

the only one that gave the machine a good defensive power. "A formation of three boats armed in this way could put up a good fight against practically any formation of E.A. encountered."[27]

The attached report by Major B.M. Jones details his views of the three proposed type of gunner nacelles tried. The following table of results was produced.

Notes on Trials of Three F.2A Flying Boats at Felixstowe

On the 13th of September I visited Felixstowe at the request of Major Hallam of the Experimental Branch there, and took part, in company with some experimental Officers from Orfordness, in trials of three Flying Boats fitted with various experimental arrangements for the gunners. The arrangements of the gunners in these 3 boats can be described as follows: (The boats are lettered A. B. and C. for reference)

A. The boat with two nacelle in the top plane. The nacelles are situated outside the engine and each is equipped with a Scarff mount which can slide fore and aft between the rear and front spars.

B. A boat with one fixed nacelle and Scarff mount in the centre of section of the top plane.

C. A boat with one nacelle with Scarff mount behind one engine. The second nacelle is to be placed behind the other engine if the scheme is adopted.

Trials Carried Out. Boat A.
This boat was attacked by 5 Scouts, and two seaters from Orfordness at a height of about 1000 feet over the sea, and a general scrap of about ten minutes ensued. I was in the right wing nacelle. Lieut. Bannister from Orfordness was in the left and a Felixstowe gunner in the front cockpit.

Boat B.
A similar fight with Captain Mitchell in the centre section nacelle.

Boat C.
Owing to engine trouble we were unable to carry out a fighting this boat, but in the evening I flew for some time in the nacelle behind the engine, and practiced moving the gun about, and aiming at clouds and other objects.

Previous Trails.
In company with Captain Musson of Orfordness I have previously engaged in a mock fight in a Standard F.2.A Boat against three machines from Orfordness. Captain Mitchell and Captain Musson had previously engaged in a similar fight on Boat A described above.

Opinions resulting from these trials.
As a result of the above trials and of previous experience with mock and real fights in land machines I hold the following opinions:

1. The Boat A, with two nacelles beyond the engine is, from the point of view of gunnery, by far the best of the three tried, and at least three times as good as the standard boat at present used.

2. The Boats B and C are a distinct improvement on the standard boats, but are not nearly so good as A.

My reasons for these opinions are as follows:

Concentration of fire.
For a large aeroplane which cannot manoeuvre rapidly it is of vital importance to be able to concentrate as many gunners as possible on any target, since the E.A. being more manoeuvrable, can as a rule choose Its direction of attack, and the only defence against it is to have superior weight of fire to counterbalance the disadvantage of the larger target presented by the larger machine.

Boat A is very much better than either of the others in this respect, although both B and C are better than the Standard boat in use at present.

Continuity of View and Fire.
In a mixed up fight against superior numbers the gunner is very handicapped if he can only command a certain portion of the sky, by the suddenness with which an E.A. appears on his field of view. In Boat A, in which the gunners on the top plane command an uninterrupted view of the upper hemisphere and can fire equally well in any direction, the ease with which it be possible to follow a complicated fight is quite extraordinary after experience in machines in which the gunner is less favourably placed. When flying low, so that there is no fear of attack from beneath, a single glance is sufficient to tell the gunner what E.A. are approaching within dangerous range, and in practice we found that even when attacked by five hostile machines we were able to keep a fairly good impression of the position of all machines near enough to be dangerous. We were all greatly impressed by the difference from this point of view between Boat A and any other machine in which we have fought. The difference of course would not be so noticeable when flying high, so that the enemy can attack from beneath.

Inter-communication between gunners and pilots.
From the forward position of the nacelles in Boat A, it is possible to see both the other gunners and the pilot at one glance, so that, with the help of the Klaxon horns

Above: Another view of the same boat at Felixstowe with dazzle painted boats in the background. There is a white outlined dark band that carries completely around the hull behind the cockpit. The identity of this machine cannot be determined from available documentation. (AHT AL0772-012)

provided for the purpose, it was possible throughout the fight to keep in close touch with the movements of the other occupants of the machine. This is of course important when encountering simultaneous attacks from several E.A. since one can tell at a glance whether anyone else is attending to a particular E.A which is becoming dangerous, or whether it is imperative to leave the E.A. at which one is firing to meet a more pressing need.

Influence of the Slipstream.
The effect of wind pressure on the gunner is very much accentuated by the slip stream from the propeller. The difference between the effect of wind pressure in and out of the slipstream is very great indeed, so much so that inexperienced gunners find considerable difficulty in aiming at all when exposed to the propeller slipstream whilst out of the slipstream very little serious inconvenience is felt at speeds below 90 miles per hour.

In boats B and C – the gunners are affected by the propeller slipstream as much as, if not more than in a Bristol Fighter. In Boat A on the other hand, there is no slipstream on the gunners, who can consequently move with much greater freedom and aim much more steadily than in either of the other boats. This point is considered to be of great Importance, especially when firing at long ranges.

Conclusions
From the point of view of gunnery either defensive or offensive the adoption of the two nacelles beyond the engines is strongly recommended.

With certain minor improvements in detail already discussed with Major Hallam, an F.2A boat equipped in this manner would be the best defended aeroplane which we have tried yet for land or sea work.

(sgd) B.M. Jones
Major, R.A.F.

Source: TNA Mun8/28/53788.

F.2A – Investigation of Blind Areas for Concentration of Fire Sept. 17, 1918

As a result of the trials Major Hallam *recommended* that the fighting tops on the upper plane at the first struts outside the engine bearers with gun rings to slide to leading or trailing edge be adopted. In making the recommendation it was considered that the boats should be developed to "their full fighting power, even if speed and range of flight are sacrificed, as the conservation of trained personnel is of prime importance."
It was recommended that:
(a) For machines (F.2A's) now on the Stations interested

(Felixstowe and Yarmouth) drawings be supplied of arrangement No.1, so that the alterations can be made locally, such parts as cannot be made on the Stations being supplied from Contractors.

(b) For F.5 arrangement No.1 be fitted for machines intended for Felixstowe and Yarmouth. The drawings for these machines (are) *now being prepared. As this will take at least five weeks it is suggested that the F.5 boats be completed as at present, and that new panels be made with the gun positions complete. These could be put in at the Stations and the old panels returned until such time as the new panels can be put into production machines.*

(c) The fighting tops to be put only into machines allocated to Felixstowe and Yarmouth.[28]

An F.5 with fighting tops and left hand propellers was taken out by Capt Scott from Felixstowe on 22 November 1918. The object was to obtain balance in the air. The machine was fitted with fighting tops mounted on top of the upper planes. "These fighting tops are of a type which can be put on or removed without altering the wing structure." The airscrews were both left handed and in order to counteract this, the tail plane was put up by ¾°.

The machine had a crew of 5 men. On the first flight one observer was in the nose, one sat by the pilot and one in the aft gun position. With this arrangement the machine was found to be slightly tail heavy. For the second flight the man in the aft position was brought forward. With this arrangement the machine flew hands off at 65 knots.

The pilot reported that the machine's flying qualities showed no variation from those usual with this type of machine.[29]

J.L. Gordon, DFC, later a Wing Cmdr in the Royal Canadian Air Force, wrote of his experiences flying F boats in the war. He considered that "three flying-boats in formation was the smallest unit capable of self-defence against enemy fighting aircraft." He noted that it was not until experienced and valuable personnel had been lost that thought was given to making better gun positions for the F boats. As far as he was aware,

the best practical gun position developed for us on twin-engined flying boats during the war was located in the top plane centre section. A hole was cut between the front and rear spar of the top plane centre section and the gun ring was mounted so as to raise it slightly above the upper surface of the top plane. A platform was erected midway between the top plane and the hull, and the gunner's legs and body protected by a stream-lined nacelle. It had a small folding door in its side through which access was gained from the hull with the assistance of a conveniently arranged rope ladder. It was possible for any of the crew to reach this position with both ease and celerity. The whole mounting resulted in a loss in speed of 2½ knots only.

It was possible from there to cover the area in rear of the flying-boat, where it was less capable of defence and from which it was most vulnerable, more efficiently than from any other position, without resorting to the doubtful practice of carrying one or two additional persons to the normal crew for gunlaying duties only.[30]

As noted above, once the idea of Fighting Tops was conceived, it was proposed that it be applied to the multi-engined bombers, the Airco D.H.10 and Handley Page V/1500, then in development.

In response to a request that the D.H.10 be fitted with "wing nacelles for fighting", the Aircraft Manufacturing Co advised that the fitting of these nacelles would reduce the top speed of the D.H.10A at 10,000 and 15,000 feet by about 12 mph to 100 mph, and reduce the ceiling by about 2,000 feet to 16,000 feet. It seemed "very doubtful whether it would be worthwhile fitting the nacelles in view of the fact that machines so fitted would not be able to fly in formation with standard machines owing in the drop in speed."[31] A further Memorandum stated that owing to the distance that the nacelles would have to be from the centre of the machine they would cause a considerable decrease on manoeuvrability.

The drop in performance of the Handley Page 'V' would be approximately 2 mph and 1,300 feet in ceiling. "Until the gun position at the back of the fuselage of this machine is found to be inefficient I do not think it worthwhile to fit the top plane nacelles."[32]

Despite this it was still planned to fit two nacelles on the top plane beyond the engines. These were to be designed, if possible, "as separate and complete units to fit on the top of the planes in such a way that they can be used or not as required." While it was considered that the Handley Page would not be fitted out in such a manner for service use, it would provide valuable information "with reference to possible improvements of future larger machines."[33] It would seem that the Handley Page referred to as being at Orfordness was an O/400 as plans of that type were requested on 31 October.

Despite the amount of work carried out no *Large America* boat went into combat fitted with the Fighting Tops. The idea was carried forward into the design of the Phoenix P.5 Cork flying boat and was to appear on the production version of the Cork, the post-war Kingston flying boat.[34] It appears that the only country to field such an arrangement during the war was the Germans who had similar type of gun mountings on their *Risenflugzeug* (Giant) bombers.

To determine the blind areas the machine is considered to be in the centre of a cube (see diagram). Each plane is the base of a square pyramid. In this method the blind areas for each gun are determined photographically and subsequently re-drawn to correspond to all six planes of vision. The blind areas are superimposed in order to obtain the combined diagram.

Assumptions were that there were three gunners in every case. The gunner in the Gallows position is assumed to apply himself to either side instantaneously to either gun as may be necessary.

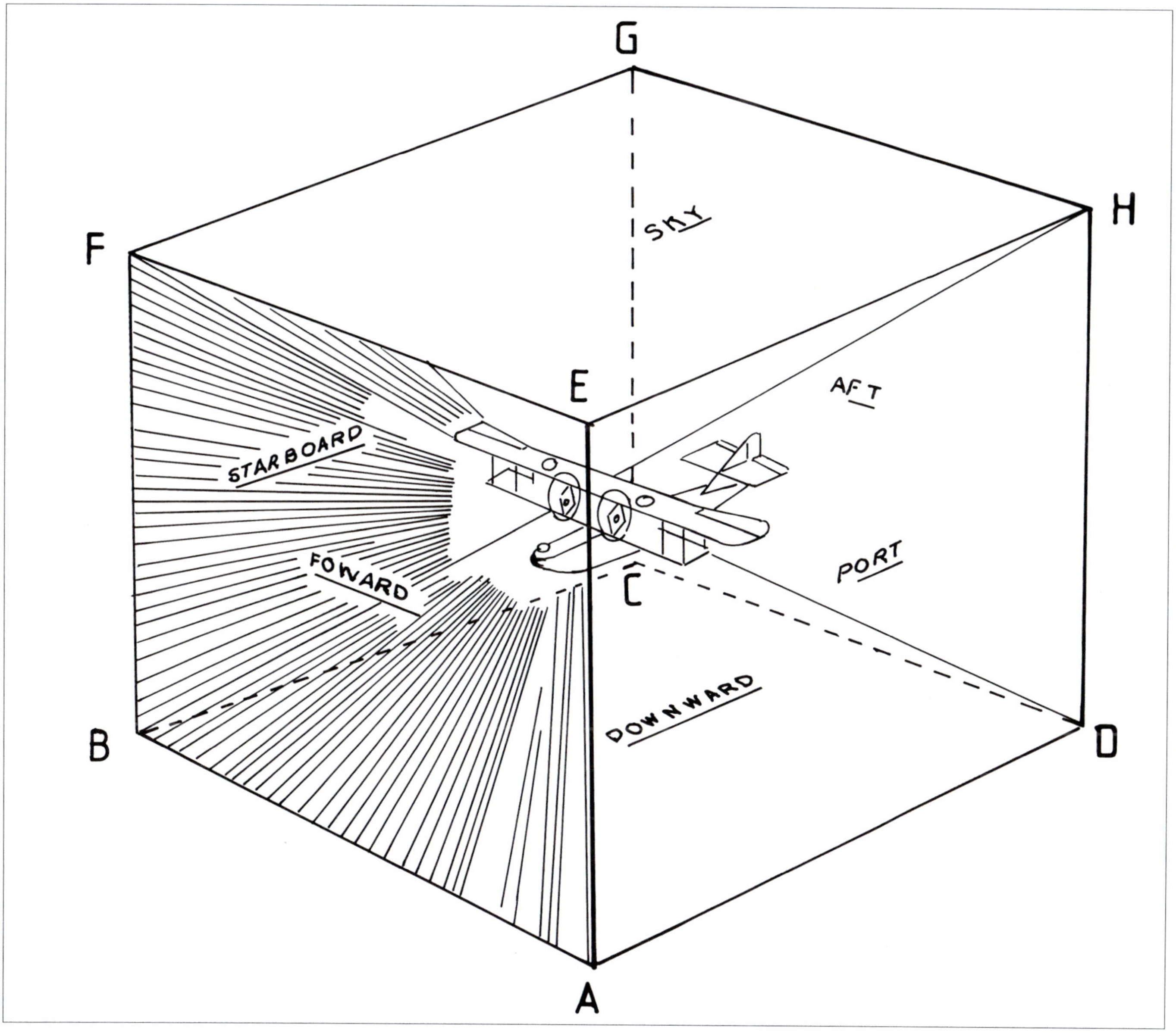

Above: Diagram of F.2A with Felixstowe Experimental Seaplane Station's Fighting Tops.

Armament of F.2A Flying Boats

1. On 5th May 1918, Major Jones and Captain Mussen visited Felixstowe Naval Air Station at the Invitation of the Commanding Officer, Lieut. Col Robertson, in order to look at the armament of the F2A Flying Boats in use there. Whilst at Felixstowe we had the opportunity of a flight in one of these boats, and flew to Orfordness, where we were attacked by a Bristol Fighter and two Camels; we were also able to try the guns by firing into the sea at a wreck and at other objects.

2. The armament with which the boat was equipped was as follows:
In the bow a single Lewis gun carried on a Scarff ring mount and equipped with a Norman wind vane sight.
For the pilot a Lewis gun on a fixed pivot.
Behind the planes were three gun positions:
A Lewis gun on the starboard side of a Gallows Mount equipped with a mechanically operating ring sight.
A Lewis gun on a type of ring mount in which the gun is mounted on a short rocking lever.
Three Lewis guns on the port side mounted together on a Gallows mount and equipped with a Norman Wind Vane sight.

Detailed Criticism of the Armament

3. Owing to its large size, the rate of manoeuvre of the boat is so slow as almost to rule out manoeuvre as an asset in fighting lighter machines. Practically the only tactics possible to the boat in the presence of well managed light E.A. are to fly close

to the water and wag the tail to prevent E.A. from staying directly behind it, and of course, if in formation, is to keep formation as closely as possible. For this reason the armament must be as quick to manage as possible, since little or no help can be given by the pilot to the gunner to keep his guns trained on the enemy. In the lighter land fighting two-seaters, such as the Bristol, the pilot flys to help his observer, and hence the observer need not always have to traverse his guns very rapidly. In these heavy machines on the other hand, the observer himself must follow the E.A. with his gun without the help of the pilot, and hence the first essential in the armament is rapidity of movement and wide arc of fire.

4. We will now consider the armament in detail in the light of the above requirements.

5. The front gun.
This is the most satisfactory gun position in the boat because the gunner is in direct touch with the pilot and has a wide arc of fire, upwards and downwards. The Scarff ring is undoubtedly the best form of mount for this position and the Norman wind vane and wing the best form of sight until such time as a satisfactory compensated sight is available for use on a ring mount.

Two guns could be used with advantage here when the new double gun Scarff ring mount can be obtained. Two guns on this mount could be easily managed because there is no slipstream and very little violent manoeuvring. Until the new double mount is available, the Scarff double gun parallel notion adaptor might be used if great care is taken to keep the rings of the mounts running freely and to balance the guns properly.

6. The Rear Upper gun.
This gun position is about the most responsible in the boat, as it is the only position from which an effective watch can be kept against surprise from the rear. The best mount for this position would be a Scarff ring; two guns could again be used with advantage. The mount that was in place in the machine in which we flew, a short pillar on a rotating ring, is very bad. A swinging pillar mount was seen upon another boat in this position, and this might have been considerably better than the other except for the fact that the swinging motion invariably jambed when moved hurriedly, and that there was not quite enough room to manage the gun quickly. We are firmly of the opinion, however, that, a Scarff double gun ring is the ideal mount for this position, and would be a great improvement over either of the mounts that we saw. As large a platform as possible should be provided for the gunner to stand on. A Norman wind vane is the best sight for a gun in this position until a mechanically compensated sight for use on a ring mount is available.

7. Side guns.
The side guns working through windows in the side are carried on Gallows mounts. These guns are certainly a great help in defending the ship because they provide the only possible means, with the present position of the crew, of fire close under the tail; they are, however, difficult and awkward to use and their only justification is the impossibility of defending the tail in any other way without radical alterations on the machine. On no account should more than one gun be used on these mounts as this enormously increases their already serious defects from the point of view of field of fire and ease of management. The mechanical sight provided with these guns is a necessity. The Norman wind vane is practically useless here on account of the disturbed air flow and because, when shooting directly backwards (the guns most useful position) the head so disturbs the air flow as to cause the windvane to wobble badly. The compensated sight having a peep and ring is the best sight to use with these guns. We can see no obvious improvement to be made in these mounts unless perhaps something could be done to allow the guns to fire more upwards without jambing.

8. Distribution of gunners.
It is understood that three and sometimes four gunners besides the pilot are available; with four gunners one would of course stand at each gun station when preparing for a fight. When three gunners only are available, one should always be in the upper rear Station, because only from here can adequate watch be kept to the rear; this gunner must be in close communication, telephonic or mechanical with the pilot, otherwise disaster is almost inevitable if attacked from front and rear simultaneously.

9. General fighting qualities of the boats.
Although the defensive powers of the machine may be considerably improved by the use of the most suitable gun mounting, we do not think that the machine in its present form can be considered in the least satisfactory as a fighter and we can see no prospect of its being able to ward off a well delivered attack from superior numbers of lighter machines manned by experienced fighting pilots. In very few directions indeed can more than one gunner bring fire to bear on an E.A. at one time, whilst the blind spot behind the tall, although to some extent defended by the side guns, is very inadequately defended, owing to the small field of fire of the guns and the difficulty of using them. The F.2.A as it stands, is, in our opinion, quite unsuitable for use as an offensive fighting machine against well managed and smaller E.A. because it can neither manoeuvre to obtain a good position, nor can it bring an overwhelming fire to bear on any enemy when the position is attained, as should be the case with a machine of such a size offering such a large target to the enemy.

Possible Fighting Qualities in a Flying Boat
10. For a heavy aeroplane of this sort to be able to put up a really good fight when out-numbered by lighter machines it is essential that as many as possible of the gunners should be able to concentrate together on all positions from which an attack is likely to be made. If this is not the case the lighter and more

F.2.A – Investigation of Blind Areas for Concentration of Fire Sept. 17,1918 *				
Type	**Present Arrangement**	**Method 1. Twin Fighting Tops**	**Method 2. Single Central Fighting Tops**	**Method 3. Single Fighting Top Behind Port Engine. F.3**
Disposition of Gunners, etc.	One gunner in front cockpit, two gunners aft with aft ring position & two Gallows mountings.	One gunner in front cockpit, two gunners in fighting tops, all with ring mountings.	One gunner in front cockpit, one gunner in central fighting top, one aft with two Gallows mountings.	One gunner in front cockpit, one gunner in fighting top, each with ring mountings, and one gunner aft with two Gallows mountings.
Percentage of sky open to three gunners	31.5	60.0	22.0	29.0
Percentage of sky open to two gunners	27.0	27.5	62.0	56.0
Percentage of sky open to one gunner	38.0	10.5	13.0	13.0
Percentage of Sky not covered by any gunner	3.5	2.0	3.0	2.0
Total	100.0	100.0	100.0	100.0

Assumptions:
1. Three gunners only in each case.
2. Limitations of field of fire due to awkwardness of handling gun mounts not considered.
3. No account taken of regions in which it is possible to see, but not fire.
4. In the case of the gunner responsible for the two Gallows positions it is assumed that he applies himself instantaneously to either gun as may be necessary.
* TNA Mun8/28/53788.

manoeuvrable E.A. will attack simultaneously from the worse defended direction, and, owing to their superior manoeuvring powers and performance can continue to do this repeatedly

11. In a boat which can come down close to the sea when outnumbered the problem of defence is simplified considerably and reduced itself into getting as many as possible of the gunners to be able to fire over as much of the upper hemisphere. Unfortunately in a sea boat the body is necessarily low in the planes and the gunners, if in the body, are therefore necessarily separated by the planes that each gunner is only able to fire over about one half a hemisphere. The defensive power of the boat for a given number of gunners for a given number of gunners in the body is therefore at least half what it would be were they all so placed as to command the whole upper hemisphere. This is because the E.A. by concentrating on either the fore or aft part of the boat need only engage the smaller half of the defending gunners at any one time.

12. For the above reason it is considered that the only way to get the best out of the available gunners is to place the two rear gunners outside the boat, level with the top planes and wide enough apart to be able to shoot straight behind the tail. If this is done no enemy can approach from in front and above without engaging three gunners at least, and none can approach from the rear without engaging two gunners, this latter area is not large enough to accommodate more than one E.A. at a time except at very long range.

13. If the development of the defence of these boats against large numbers of smaller E.A. is intended to be carried beyond simply improving the armament on the existing boat as it stands, we think therefore, that the best direction for improvement consists in making it possible for the two rear gunners to fire from Scarff mounts places in the rear trailing edge of the top wings, either just behind the engines, or further out than the engines. These gunners would then be equipped with double mounts, and the armament of the boat would consist of six guns capable of concentrating forward and upward, four guns being available for almost the whole rear hemisphere, and no blind whatever except at the very short range. Such a machine it is believed would put up a very good fight against quite heavy odds in the form of numerous small machines, and would be between two or three times as effective

as the present machines with the same number of guns and gunners. The unsatisfactory gallows mounted guns at the sides be guarded by the two gunners on the top plane.

14. With regard to the question discussed at Air Technical Committee Meeting on 30th May 1918 of making a long distance three-seater fighting Flying Boat, we are now definitely of the opinion that no sea boat of the size required is likely to be sufficiently manoeuvrable for its manoeuvring powers to be of serious value in a fight with lighter craft, and therefore that, if such a boat is to be fitted against lighter craft, it must contain four occupants, i.e. three gunners distributed as in paragraph 12. In other words the only alterations for this type of work are, a boat armed so as not to have to rely on manoeuvre at all, and a single or two-seater machine carried in some way, close to the spot at which it is to be used. The long-distance three-seater fighting boat is definitely considered to be an unsatisfactory proposition and unlikely to repay experiment.

Source: TNA MUN8/28/53788

Above: The Brandenburg W.19 was a larger version of the W.12 and mixed patrols of the various types of Brandenburgs were carried out. Its greater power enabled it to carry two fixed guns for the pilot as well as a flexible observer's gun plus more fuel for longer range and greater endurance.

Chapter 8 Endnotes

1. Memorandum for Captain Mustin, Subject: R.A.F. Experimental Station, Isle of Grain. 10.07.1919. J.L. Callan Papers Library of Congress Manuscript section.
2. *Aircraft v. Submarine. Submarine Campaign 1918*, Vol.1, Part 4. Of the *Technical History and Index*, a Serial history of Technical Problems dealt with by Admiralty Departments, Technical History Section, March 1919.
3. Minutes: Construction Progress Committee Meetings of 05 and 26.06.17. TNA AIR1/270/15/226/113.
4. Admiralty to Vice Admiral, Dover Patrol. Signed A. Flint for Secretary. 22.02.1918. TNA AIR1/643/17/122/267.
5. Capt Lambe, Dunkirque, to Vice Admiral, Dover Patrol, 27.02.1918. TNA AIR1/643/17/122/267. Note: Author's italics.
6. Note to M Branch by W.A. Medrow. 20.08.1918. TNA AIR1/643/17/122/267.
7. Wing Commander J. Porte, 17.03.1918. TNA AIR1/643/17/122/269. The cover sheet to this report is signed by a Rear Admiral and he states that he was in complete agreement with Porte's views.
8. Minutes, Air Department Progress Committee, 29.05.1917. TNA AIR1/270/15/226/113.
9. For the story of the Brandenburg floatplanes see Owers, C.A. *Brandenburg Aircraft of WWI*, Vol.2 and 3, Aeronaut Books, US, 2015.
10. The NAF referred to these as "Outrigger gun mounts".
11. Report to Controller, Technical Dept., on Trials of Gun Positions on Flying Boats. 17.09.1918. TNA MUN8/28.
12. *Attack on F.2A by 5 Monoplane Seaplanes.* No.4 (Operations) Group Bulletin, July 1918. TNA AIR1/288/15/226/138.
13. Report to Controller, Technical Dept., on Trials of Gun Positions on Flying Boats. 17.09.1918. TNA MUN8/28.
14. Director of Air Dividsion, 24.03.1918. A.D.351. TNA AIR1/643/17/122/269.
15. For the story of the Caudron R.11 see Owers, C.A., *Caudron R.11*, Datafile No.161, Albatros Productions, UK.
16. Forty-Fifth Meeting of the Technical Committee, 30.05.1918. Copy in US NARA RG72 Bureau of Aeronautics, General Correspondence initiated in the Bureau of Construction & Repair 1917-1925, Box 425, File O-ZS-7. In a parallel development the USN in France suggested in June 1918, that a "fighting seaplane with nearly the same fuel range as the patrol machine, but carrying no military load and no more crew than is absolutely necessary for an efficient fighter" be developed.

A twin engined tractor flying boat with a crew of three, comprising a pilot and two gunners, one forward and one aft, was proposed. The matter was to be placed before the CO of Killingholme for his recommendation for this type of machine for operation with the patrol seaplanes from that station. It appears that the Navy was aware of the development of the Curtiss HA Dunkirk fighter and it was believed that this machine could be modified to suit the requirements. The HA was a single engine floatplane without the endurance to fit the requirements and was eventually to prove unsuccessful. (See Owers, C.A. "The Curtiss HA Dunkirk Fighter", *Over the Front* Journal, Vol.28, No.3, P.230.

17. Letter: Col Samson, Commanding No.4 Group SE Area RAF, Felixstowe, to R-Adm Hickley. 23.07.1918. TNA AIR1/643.
18. Rear Adm Hickley to Admiralty, *Felixstowe Air Station – Provision of flights of fighting seaplanes*. 23.07.1918. TNA AIR1/643.
19. Minutes of 10th Air Department Weekly Progress Meeting, 17.07.1917. TNA AIR1/270/15/226/113.
20. Admiralty to Air Ministry. 30.07.1918. TNA AIR1/643.
21. Letter: Air Ministry to Admiralty, *Flying Boats – Protection of from Enemy Aircraft*, 29.08.1918. TNA AIR1/643.
22. From Armament Experimental Station, Orfordness, to Assistant Controller (Experiments) Technical Dept, Aircraft Production. 28.08.1918. TNA AIR 2/738.
23. Report: *Mock Fight, 22.8.18, between a Felixstowe H12 Flying Boat with top plane nacelles and 2 scouts and 2 two Seaters from Orfordness, viz – Dolphin, Camel, Bristol Fighter and Bulldog (230 Clerget)*. TNA AIR 2/738
24. Memo from Assistant Controller (Experiments), 19.09.1918. TNA AIR2/732/284/59.
25. Report on Comparative Trial of Various Types of Gunners Nacelles on Flying Boats on 13.09.1918 at Felixstowe. Copy in TNA Mun8/28.
26. The serials were recorded without their N prefix. N4339 was an H-12 Convert, not an F.2A.
27. Report on "Comparative Trial of Various Types of Gunners nacelles on Flying Boats on 13.09.1918 at Felixstowe." TNA Mun8/28/53788.
28. Major Hallam's "Report to Controller, Technical Dept, on Trials of Gun Positions on Flying Boats." 17.09.1918. Copy in TNA Mun8/28.
29. Test Report No.18, Seaplane Experimental Station Felixstowe. "F.5 – Preliminary trials with fighting tops and left hand propellers." TNA AIR2/896.
30. "War Experiences (1914–1918)" by Wing Cmdr J.L. Gordon, DFC, Royal Canadian Air Force. Source not stated but thought to be from the series of Lectures and Essays from the Work of the Officers attending courses at the Royal Air Force Staff College, and published by the Air Ministry. Photocopy in RAFM J.M. Bruce Collection Box 61.
31. AMC to Controller, Technical Dept, Aircraft Production. For Attention of Col Ogilvie. 07.10.1918. TNA AIR2/738.
32. Memorandum: *Top Gun Nacelles*, from Assistant Controller (Design). 09.10.1918. TNA AIR2/738.
33. Letter: Controller, (Experimental) to OC Armament Experimental Station. Orfordness. 26.10.1918. TNA AIR2/738.
34. See 'Owers, C. "The Phoenix P.5." *Cross & Cockade International*, Vol.45, No.1, P.1.

Above: Just as the W.19 was a larger, more powerful development of the W.12 biplane, the Brandenburg W.33 was a larger, more powerful development of the W.29 monoplane that could carry two fixed guns for the pilot as well as a flexible observer's gun and more fuel for longer range and greater endurance. A small number reached the front before the Armistice and it had a long, successful postwar career.

9. Camouflage & Markings

Above: H-12 867X displays the two toned colour scheme seen on many of the early H-12 boats. Despite Sir Austin Robinson's recollections that all the H-12 hulls were grey above and blue grey below, from the tonal values in photographs, the upper colour appears to be the Curtiss British Khaki, while the lower light colour could well be a bluish grey. The serials appear to be white while fabric surfaces appear to be clear doped. The rudder stripes end at the ply covering that protected the bottom of the rudder that would enter the water. Slow landing the H-12 meant the tail hit first before the step, this was a great defect of the H-12.

Notes on markings are given in the captions to the photographs throughout this book. Where there is more than one photograph of a particular machine, these photographs have been kept together in the main to show how lighting variations can change the interpretation of colour in black and white photographs.

The Air Department of the Admiralty issued Temporary Memorandum No. 48 in 1915.[1] This is the earliest document that covers the application of dope to British marine aircraft that the author has so far discovered. The Memorandum was issued due to "certain firms having been found employing unsuitable dope on the fabric of aeroplanes and seaplanes." All firms were ordered to use one of the three following dopes in future:

1. Cellon Dope.
2. Emaillite Dope.
3. Bi-nitro Cellulose Dope (as used by Messrs. Short Brothers).

In the case of Cellon and Emaillite the dope was to be applied in the following manner:

First coat: 2 parts dope, 1 part acetone.

Second coat: 4 parts dope, 1 part acetone, then 4 coats or if necessary 6, of plain dope.

After the dope had thoroughly dried varnish was to be employed to make the wing waterproof, or else the following solution was to be used:

Take one part of best boat varnish and mix with two parts of petrol. Spread this over the wings with a brush, but brushing must not continue when the varnish becomes sticky.

Technical Memorandum No.59 dated 31 July 1916, on the "Use of Dopes" provides more information on the RNAS doping practices. This was a general instruction following the issuing of the instructions to all firms now building aircraft for the Admiralty.

Only the following non-poisonous dopes were to be employed in future:
(1). For use on machines of 1st class military importance -
Emaillite, No.11a and 11c.
Cellon, No. N.P.2.
Raftite.

Above: N4081 displays the "C" for Calshot station on its fin. The gap between the cockpit enclosure and the hull is well shown by the crewman standing in the opening. The dark colour of the hull, fins and washboard is thought to be a black anti-fouling bituminous paint. Note the position of the serial on the rudder. (AHT AL0060-121)

(2). For School Machines-
Emaillite No.11c.
Cellon, No. N.P.2m made with Acetone substitutes.

The poisonous dopes containing tetrachlorethene are not to be employed in future unless the above non-poisonous dopes are unobtainable.

All machines were to be varnished on their upper surfaces with pigment varnish P.C.10, and on their under surfaces with varnish V.114. On no account was V.114 to be used on upper surfaces. If P.C.10 was not available then ripolin could be used instead.

Little documentary evidence appears to have survived about the doping and marking practices of the RNAS/RAF with regards to flying boats. The following is the best estimate given the present state of research into this topic. Much remains to be done on this neglected subject.

J.D. Rennie gave the following description of the finish applied to the *America* flying boats in a lecture to the Royal Aeronautical Society in 1923. Considering Rennie's association with Porte this may be taken as the practice that was adopted as standard. It is considered that this was applied to the Felixstowe constructed boats due to his work at Felixstowe. Whether it applied to the earlier AMC series of *Small America* boats is unknown. The H-12 boats were delivered in Curtiss colours. What instructions Curtiss received as to how to finish the boats is unknown at this time. It is probable that this scheme was applied to the hulls of the H-12 boats as it was intended to allow for better hydrodynamic properties and not for camouflage purposes. Careful consideration of photographs will reveal if the hull was treated in this fashion.

Rennie recorded that the inside as far as the fin top and the end of the tail was coated with bitumastic paint, while the remaining portions were well varnished. The bottom outside was painted with a special flying-boat black anti-fouling composition. The first hulls constructed had their bottoms varnished, and when the bottom was still tacky with their third coat of varnish, black lead was rubbed in and the surface polished. This treatment gave excellent results but required a lot of work at the stations to keep the bottoms in good condition. Black lead was applied for the reason that when the bottoms were nearly dry it could be polished to the smoothness of glass, thereby assisting the hull get off the water at high speeds.

This treatment was succeeded by black anti-fouling paint. This could cause problems as the composition required stirring all the time and the boat had to be placed back in the water while the anti-fouling was not properly dry.[2]

The outside of the hull that did not require painting, such as the nose planking, hatches, top of fins, three-ply work on the top sides, etc., received three coats of varnish. Steel fittings were given two coats of black stove enamel.[3] This finish can be seen in period photographs of British H-12 Convert, F.2A, F.3, F.5 and H-16 seaplanes.

H-4

To date no documents have been discovered that describe the finish applied to the *Small America* boats. It is probable that fabric covered areas were doped and then clear varnished while hulls were also varnished.

H-12

The Curtiss Company started to use Valentine's English Khaki when it began to produce machines for the British. The

Above: A late model H-12 with Calshot's "C" on the fin. The top surfaces of the wings and the tail are a dark colour. The hull appears to be in the two-colour scheme but this may be due to the lighting conditions. Other Calshot H-12 boats had the "C" carried on the top wing on both sides of the centreline over the inner pair of interplane struts, and it is likely that this was standard practice for Calshot boats. Note the bomb under the wing and how many personnel were required to handle these big flying boats.

description of "khaki" is misleading as the colour was actually a light tan colour. It is thought that this was the upper colour that was applied to the first batch of H-12 boats. According to Sir Austin Robinson the H-12 hulls were "always painted grey; in some cases the fin surfaces and the lower part of the after body were painted, yacht-wise, in a darker blue-grey shade."[4] 8651 bore a two colour scheme on its hull while 8666's hull was a light colour (grey) overall.

After much discussion as to commercial secrets, economic advantage, etc., the Valentine Company gave the US Navy the formula for their Velspar Curtiss English Khaki Gray Enamel as follows[5]:

French Washed Ochre	19.0%
Leaded Zinc Oxide	20.0%
Burnt Turkey Umber	3.0%
Pure Bone Black	0.5%
Chrome Green (Lead Chromate & Prussian Blue)	0.5 %
Valentine's Grinding Valspar No.1	12.0%
Grind add	
Valentine's Grinding Valspar No.2	45.%
	====
	100%

In August 1917 it was reported that problems had occurred with the fabric of *Large America* boats after mooring out. The top planes suffered the worse and it was definitely due to the sun and not to dampness. These planes were doped with American dope.[6] A further report in October recorded that on the *Large Americas* at Scapa the dope had worn badly except where the identification marks were painted, this suggesting that the wings should be painted all over. The wings giving trouble had been received from America. While this would add considerably to the weight of the wings it was noted that English wings doped with Emalite and P.C.10 would be perfectly satisfactory. There is no record of this being adopted, however P.C.10 was adopted by the RFC/RNAS to protect fabric from the effects of the sun's rays and was in plentiful supply. The green colour stated to have been applied to British seaplanes could not have been P.C.10 as this was a brown colour.[7]

F.2A, F.3 and F.5

The late C. Rupert Moore published his findings on British colour schemes applied to the *Large America* boats.[8] Moore had the advice of Sir Austin Robinson who flew these boats and whose research into the Felixstowe boats is reflected in the main text Endnotes.

According to Moore the fabric covered surfaces of the F.2A and later boats were given several coats of dope followed by

Above: N4566 from May, Harding and May, in what appears to have been a "standard" scheme for the *Large America* boats late in the war. The airscrews have canvas covers. The serial, carried above the tailplane on the rudder, can just be made out on the original photograph. This boat was at Calshot from 1919 to around 1924.

several coats of a marine varnish. This gave a glossy finish that imparted a slightly yellowish tinge to the underlying fabric, in time as it weathered to a dull surface. The upper surfaces were given a coat of dark green that Moore describes as an early form of Nivo. To date no documentation has been found to confirm this.

Engines were left in natural finish with fuel, oil, radiators being painted Battleship Grey. In fact all the odd and ends such as sockets, windmill pumps, etc., were painted in Battleship Grey.

Despite this, it is evident from photographs that the fins, hull bottoms and washboard were painted a dark colour, the black bituminous anti-fouling paint as described by Rennie.[9] The interior of the hulls was clear vanished and this is evident in the photographs in this volume. The *R.A.F. Flying Manual Part III Seaplane Flying* was published in 1926 but would still reflect the practices of 1918. This publication notes that below the water line, both inside and out, the woodwork of hulls should be frequently treated with a thin coat of varnish or anti-fouling paint laid on a smoothly as possible.

Cockades were the usual vermillion, white and ultramarine and glossy when applied. The location varied and reference to photographs should be made as early machines carried cockades on the upper wings only, but later they were carried on the upper surface of the top wing and on the bottom surface of the lower wing. Hull cockades were not always applied and varied in position.

The various contractors that built the Felixstowe hulls appear to have finished them in different interpretations of the official instructions that must have been issued. According to Robinson the vertical fabric work of May Harding May was usually olive green; that of the early Saunders was usually natural fabric colour."[10] Saunders hulls were varnished to a gloss finish on their sides with the fabric rear section left clear doped but later were doped the dark pigmented colour the same as the wings (green or P.C.10) as may be seen in the accompanying photographs. The small lots that contracts were let for would have contributed to the profusion of variations in the colour schemes applied to these flying boats.

A Report in July 1920 noted that the fabric on Curtiss H-16 N4892 was "still rum taut and is in excellent condition." This was from a boat that was delivered to Felixstowe for erection in the Week Ending 25 May 1918, and first flown on 15 August. The aircraft had since done 60 flying hours, and had been moored out for considerable periods in bad weather. A pair of H-16 wings that had been stored for two years were also in excellent condition. It was considered that the "Curtiss system of Doping fabric appears by results to be far superior to any system used on British built machines."[11] Curtiss had improved a lot since the reports of August 1917.

The colours of USN boats both in the US and in the war zone was Navy Grey overall. The Bureau No. was marked on the white central rudder stripe in black characters. On aircraft in the US it was also carried on the hull in large white characters. The seaplanes in France and England usually carried a station marking with an individual aircraft number.

Dazzle painting:[12]

On 4 June 1918, occurred the largest encounter between the fighting flying boats and their German adversaries. During this encounter the five British flying boats found that they lost sight of their companions in the melee or were forced down, and would form up on a boat in what they thought was their position in the formation only to discover that they had the

Above: Close-up of a late F.2A that appears to have a Saunders' hull. The nose and, fins and washboard are a dark (black?) colour while the ply forward sides of the hull are clear varnished. The rear and top decking fabric is clear doped (with the fabric top decking possibly white). The wings are doped a dark colour on their upper surfaces. The struts appear to be Battleship Grey. Note the prominent footsteps up the side of the hull.

Above, Below, & Bottom of Facing Page: A May, Harding and May boat out of an order for ten F.2A patrol boats (N4510-N4519), N4510 is seen in another scheme applied on many boats. Note the parallel front tailplane struts. The hull fabric appears to be a dark colour (P.C.10 or green?) the same as that applied to the wings. The hull wood appears to be varnished except for the nose that is much darker (black?). The fins, washboards and hull bottom are probably the same dark colour. This seaplane had a short career. After being delivered to Felixstowe for type trials in January 1918, it was with the Felixstowe patrol by March, but an emergency alighting due to engine failure saw the boat sunk while under tow in heavy seas on 8 April.

Above: F.2A N4516 was from the same batch as N45106 and in the same colour scheme. Different light changes the tonal values of the photograph such that in this photograph the fins appear to be the same colour as the rear fuselage while the washboard looks darker. Note the Short seaplane with folded wings in the background. N4516 was delivered to Killingholme in February 1918. It had a chequered career breaking its tail when launching on 3 June; having the elevator shot away on 19 July, and dropping two bombs on a U-Boat off Scarborough on 23 July. At this time it was with the USN. It was written off in a crash off Immingham on 14 August 1918, without injury to the crew.

wrong boat. As C.F. Snowden Gamble recorded:

With the exception of Hodson's machine (which 'was terrible in appearance , painted post box red, with yellow lightning marks running across her' – for he 'fondly hoped that this would put the wind up the Hun'), all had plain varnished hulls, so that it was difficult to distinguish one from the other."[13]

As a result of this action

it was decided that all flying boats should have their hulls 'dazzle-painted', primarily with the object of being able to distinguish them clearly in the air, so that a pilot could, in a 'dog-fight', know at a glance who was in a particular machine – knowledge frequently of considerable value.

Pilots were allowed to paint their machines as they desired, and a result the bizarre was not lacking.[14]

Later G.E. Livock claimed credit for him and Bob Leckie starting the colourful dazzle painting of the *Large America* flying boats, as noted below. Very few flying boats were given a Dazzle paint scheme and these were those associated with Felixstowe and Great Yarmouth, the stations whose boats would be engaged with the German fighter floatplanes. According to Thetford the scheme of each flying boat was charted and these charts were distributed to all air and naval units operating off the East Coast. Felixstowe imposed a standard to the painting of the flying boats in that they were to be in a scheme of stripes and squares. Yarmouth was where the more bizarre schemes originated where the pilots had more latitude in their choice of markings.[15]

The colours applied were the same as those for the cockades with the addition of green, bright yellow and black. It seems probable that other stations would have followed on with their own variations of the dazzle paint schemes, but again documentation is lacking.

The aircraft illustrated show some of the dazzle schemes adopted and their colours where known.

F.2A N4287 was black overall and named *Black Bess*, while N4291 was also doped black in late 1918.

N4289 was painted post box red with yellow lightning marks running diagonally across.

N4283 was *Bob* Leckie and G.E. Livock's F.2A and was painted dazzle painted in "a wonderful black and white

Above: Another F.2A from the same batch N4515 has the rear hull fabric clear doped while the top decking appears to be the same colour as the fins and forward ply areas of the hull. These photographs show the difficulties in trying to determine colours without documentation on the various colour schemes applied to these boats. It must be remembered that the coloured dope was not to apply camouflage but to protect the fabric from deterioration from the action of the sun's rays. Note the Sopwith Camel on a lighter to the right of the seaplane. The deck was canted up so that when under tow the deck was approximately horizontal to launch the fighter. (via J.D. Oughton)

striped design, thereby starting a fashion. Before long all our boats looked rather gay and rather fearsome in colour schemes chosen by the individual crews."[16]

In the Imperial War Meuseum's collection is a painting by C.R. Fleming-Williams showing three *Large Americas* attacking a submarine. One is shown with a "natural" hull colour while the other two have dazzle painted hulls in red and blue over white.[17] In 1936 the first impression of *The Clouds Remember* by Oliver Stewart, contained a painting by Leonard Bridgeman of an F.2A in red and white checkerboard pattern to the front of the hull and red and white stripes to the rear hull. Given these gentlemen's association with aviation, the colours are thought to be authentic.

Post-war the USN adopted a similar scheme for its F-5-L flying boats, and for the same reasons. It appears that the Atlantic Fleet Air Detachment's Airboats had a system of black and white stripes/squares to enable each aircraft to be identified in joining up for squadron formations. When this detachment made the first flight to Panama in January 1921 the upper wing was painted in black and white stripes to aid in recovery if forced to alight at sea. In December 1922 the F-5-L boats of Scouting Plane Squadron One were allowed to be painted as follows:

Division One: White markings over Navy Grey hull.
Division Two: Yellow over blue hull.
Division Three: Black over yellow hull.

The F-5-L boats of the Pacific Air Detachment were painted in accordance with regulations with the upper surface of the top wing and the skid fins in yellow.

British doping practices post-war are indicated in the report of the British Aviation Mission to the Imperial Japanese Navy by The Master of Sempill who commanded the British Mission. It was decided to use P.D.N.12, a pigmented nitro-cellulose dope developed at the RAE. Three coats being used. In addition nitro covering V.84 aluminium was used, two coats being applied over the P.D.N.12.[18]

Endnotes:

1. RNAS Memoranda 1915, Air Department No.6, January 1916. TNA AIR1/2097/207/15/12.
2. Further details are inferred in a letter from the French flying boat manufacturer of Hydroavions J. Donnet wrote to the Commander of the RNAS Station at Malta enquiring about the way their Porte boat bottoms were made watertight. He wrote that officers had informed him that the "Porte Boats are tarred" and he asked if the description of the ingredients and a small sample could be provided as they were having trouble with the three-ply that they used in their hulls. Malta was given permission to pass on the information but unfortunately that letter is missing from the file. TNA AIR 2/989.
3. Rennie, J.D. "Some Notes on the Design, Construction and Operation of Flying Boats," Paper delivered to the RAeS on 18.01.1923. Paper published in the *Journal of the RAeS,* 1923. P.123. These instructions for finishing a hull are repeated by B. Robinson in *Aircraft Camouflage and Markings 1907-1954,* (3rd Revised impression), Harleyford Publications

Above: N4287 was a Saunders hull and appears to have been reconditioned with a larger and unusual style of serial. N4291 had its serial painted in a similar style and it is possible that this was done while the two were in USN service at Killingholme. The front of the hull, washboard and fins appear to be grey with the fabric surfaces a dark colour (green or P.C.10?). The style of cockade is also of interest. Note the side window behind the cockpit canopy, a feature lacking on the H-16. This seaplane was in use at Killingholme and was one of those handed over to the USN. It attacked U-Boats on 26 and 28 June, and on 13 July 1918, while being flown by USN crews. It was written off in the W/E 5 September 1918.

Ltd, Herts, 1959. Apart from this there is little archival information available with regard to the finish of British flying boats.

4. Robinson, A. "The Recognition of First War Flying Boats", *Cross & Cockade Great Britain,* Journal, Vol.11, No.1, 1980, P.21.
5. Valentine Co, 456 Fourth Ave, NY, to Bureau C&R. 17.02.1917. NARA RG72 Box 72, File 13-10.
6. Minutes, Air Department Progress Committee, 28.08.1917. TNA AIR1/270/15/226/113.
7. The late J.M. Bruce recorded the actual colours of P.C.10 in his records held by the RAF Museum and he was adamant that the colour of P.C.10 was brown.
8. C. Rupert Moore, "Felixstowe F2A Flying Boats," *Scale Models*, UK, December 1977, P.616.
9. *Felixstowe F.2A Early* and *Felixstowe F.2A. Late,* Instruction booklets for the Wingnut Wings kits. These instruction booklets contain a wealth of information gained in the Company's researches into the aircraft that they choose to model. The booklets contain detail photographs and are well worth downloading from the Wingnut Wings website.
10. Robinson, A. *Op Cit.*
11. M&AEE, Isle of Grain Report N.F. Report 255, 05.07.1920.
12. Norman Macmillan wrote extensively for aviation periodicals and journals as well as a number of books on his experiences post-WWI. He noted that Dazzle Painting was invented by the marine artist Norman Wilkinson for the camouflage of merchant ships and wrote the "use of dazzle painting on some British flying boats during the war of 1914–18 was the only application of this system of camouflage to aircraft." He apparently was under the mistaken impression that the dazzle painting was to conceal the boats whereas it was designed to make the boats stand out to their comrades in an action. "Floatplane and Flying Boats" by N. Macmillan, in Winchester, C. (Ed), W*onders of World Aviation,* UK, circa 1937. P.906.

Above: An F.2A in flight showing the dark bottom of the hull.

Above: N4402 with dark rear to the hull. It would be extremely interesting to find the specifications that Phoenix worked to in finishing its Felixstowe flying boats. This machine is on the Brough slipway and is in the factory finish before allocation to a RNAS/RAF unit.

13. Snowden Gamble, C.F. *The Story of a North Sea Air Station.*
14. *bid.* P.400.
15. Thetford, O. *British Naval Aircraft Since 1912,* Putnam, UK, 1982. P.199.
16. Livock, G.E. *To the Ends of the Air,* HMSO, London, 1973, P.61.
17. This painting is reproduced in King, B. *Royal Naval Air Service 1912–1918,* Hokoki, UK, 1997.
18. "The British Aviation Mission to the Imperial Japanese Navy" was an extract from a paper given to the Royal Aeronautical Society on 02.04.1924, and published in *Flight* 10.04.1924, P.209. A similar scheme was adopted by the RAAF for its Fairey IIID floatplanes and is considered would have been that applied to the seaplanes of the RAF in this period.

Above: N4297 in factory finish. This machine appears to have the forward fuselage, washboard, fins and hull bottom in the same dark colour as the upper surface of the wings and tailplane. The hull top decking panels appear to be clear doped as is the rear hull fabric. Note the light (white?) panel where the hull cockade would be applied. The wing cockades overlap the ailerons. N4297 was later given a dazzle colour scheme. Note the additional Lewis Gun at the cockpit for use by the 2nd pilot.

Below & Facing Page: N4400 was the first machine off the third and last order for F.3 boats from the Phoenix Dynamo Manufacturing Co. Note the style of serial application. The Phoenix c/n is stencilled under the tailplane. These factory photographs are presumed to be taken at the official launch of the boat with factory and RNAS personnel in attendance. Compare the details of this boat with the photographs of F.3 boats from earlier batches.

Felixstowe F.2A N4296 assigned to Felixstowe.

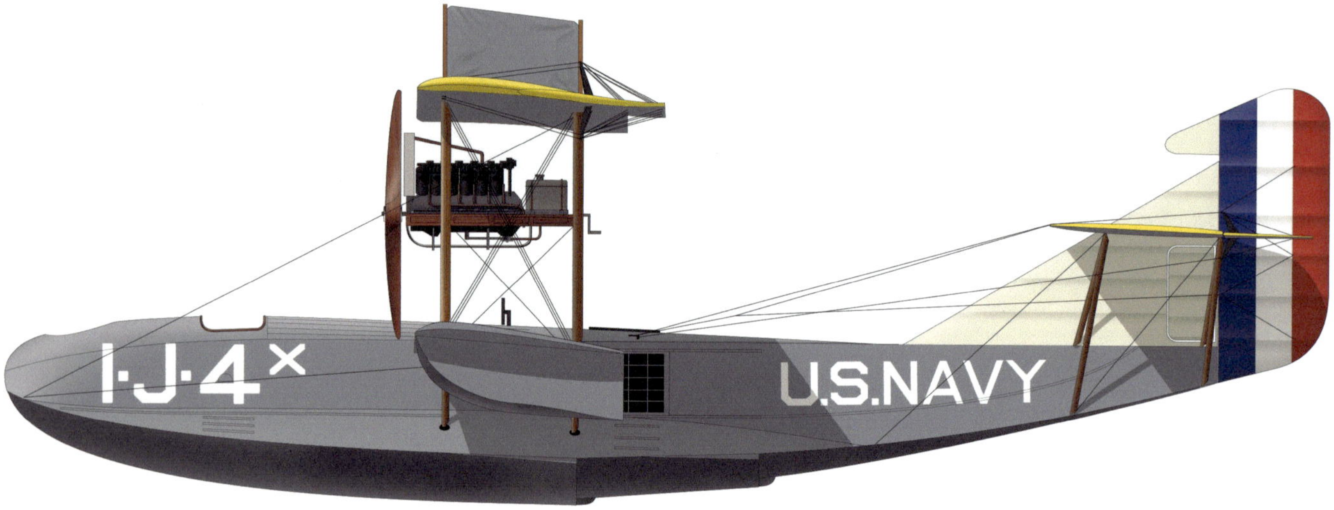

F-5-L assigned to Utility Sqdn 1, U.S. Navy.

F-5-L A4293, U.S. Navy. The finish appears to be aluminum overall except for the bottoms of hull and floats.

Above: N4401, by contrast to N4400, is in a light finish to the hull. N4400 was delivered to Killingholme in March 1918, while N4401 was delivered in April. 4401 was written off in June 1918 after being wrecked. It could be that even numbered hulls were given the dark finish and odd hulls the light finish, but not enough photographs have been examined to prove or disprove this theory.

Above: N4413 also is in a light finish to the hull, but the fins and forward fuselage has the black lines applied. On this machine the serial is not as elaborate as on N4400, but there is a more elaborate application of the c/n under the tailplane. It reads: "PHOENIX BRADFORD, No. 213." Brough slipway, 27 June 1918.

N4401

PHŒNIX BRADFORD
Nº201
N4401

Above & Facing Page: Views of N4401. The light colour of the top of the fins compared with the washboard and hull bottom is well displayed in the close-up of the rear hull. It appears that the forward hull and top of the fins were clear varnished with the washboard and hull nose and bottom black. The hull top fabric panels were clear doped. This clear doping of the top panels may have been to allow for light to penetrate and this can be noted in the construction photographs contained in this volume.

Short F.5 No.7 of the Japanese Naval Air Service, 1924. Japan received 10 Short-built F.5s in 1921 and later produced the type under license, at least 50 being built by Aichi. In 1929 16 were reported in front-line service with a further six in reserve.

Above, Below, & Facing Page, Top: Three photographs taken of N4297 of No.230 Squadron from an accompanying boat. N4297 was another Saunders hull. The cockpit cabin has been removed. The boat has been described as being painted with blue and white stripes on the hull forward and in bands to the rear, however this has not been confirmed. Note the three white bars on the upper surface of the top plane. Delivered to Felixstowe in May 1918, this machine was engaged, along with N4513 and N4540, in a fight with Brandenburg floatplanes on 4 July. Although forced down the seaplane was able to take off and return to Felixstowe. It was still at the station in January 1919. (AHT AL0772-007)

Above: N4283 in the black and white scheme that Leckie and Livock applied to their F.2A in late March 1918, some months before the general introduction of the dazzle schemes. N4283 attacked a U-Boat on 17 March in company with N5295. This boat was still on the strength of Great Yarmouth in January 1919.

Above: A poor photograph of an F.2A with canopy fitted to the cockpit. The dazzle scheme appears to be that of N4283.

Above: F.2A N4083 on board a lighter H-19 at Fleixstowe. The hull is criss-crossed by the dark colour to form a diamond pattern with the light colour. Colours quoted as white over dark green. Note the serial carried low on the rudder. N4083 was delivered to Felixstowe on 22 August 1918, and was still there in January 1919.

Above: N4490 was named *Aquila* and the dazzle scheme was not carried under the hull, nor are the wing tip floats in the dazzle scheme.

Above: N4258 with the front of the hull dazzle painted. (via R.W. Elliott)

Above: N4258 after the rear of the hull was included in the dazzle scheme. (via R.W. Elliott)

Below: This F-boat has also been identified as N4258. Note the addition of F.3 to the boat's bow.

Above: This F-boat bears the same scheme as N4258 and has the F.3 marked on the bow. The scene is outside the camouflaged No.3 Hangar at Felixstowe. (AHT AL0772-015)

Below: N4099 has only the rear of the hull dazzle painted. Note the wing cockade has not been painted on the replacement balanced aileron. (via J.D. Oughton)

Above: Close-up of nose of a dazzle painted late F.2A. "Crashed one week later."

Below: British H-16 with zig-zag stripe to forward hull and light coloured nose. Note how the dark upper wing colour is carried around the leading edge of the wings.

Above: Dazzle painted H-16 N4060 "ready for patrol" at Felixstowe. (AHT AL0772-017)

Below: N4060 beached. Note the balanced rudder and ailerons. The under wing cockade does not carry over onto the aileron. Another dazzle painted boat is on the water behind N4060. This would indicate that the balanced ailerons were added in the UK.

Above: Close up of the rear hull of a dazzle painted Felixstowe boat. On the original three colours can be made out in the dazzle pattern. Note the mounting for the machine gun being demonstrated on the upper hull position. (Nicolas Cooper via SEAWINGS)

Below: N4304 has stripes from the bow to the stern of the hull. Note how the wingtip floats were painted the same as the hulls.

Above: The dazzle painting was carried over onto the hulls of the F-boats. This photograph is said to have been taken at Felixstowe. This example has the cockades under both upper and lower wings. Note the crowded beach; a post-war photograph?

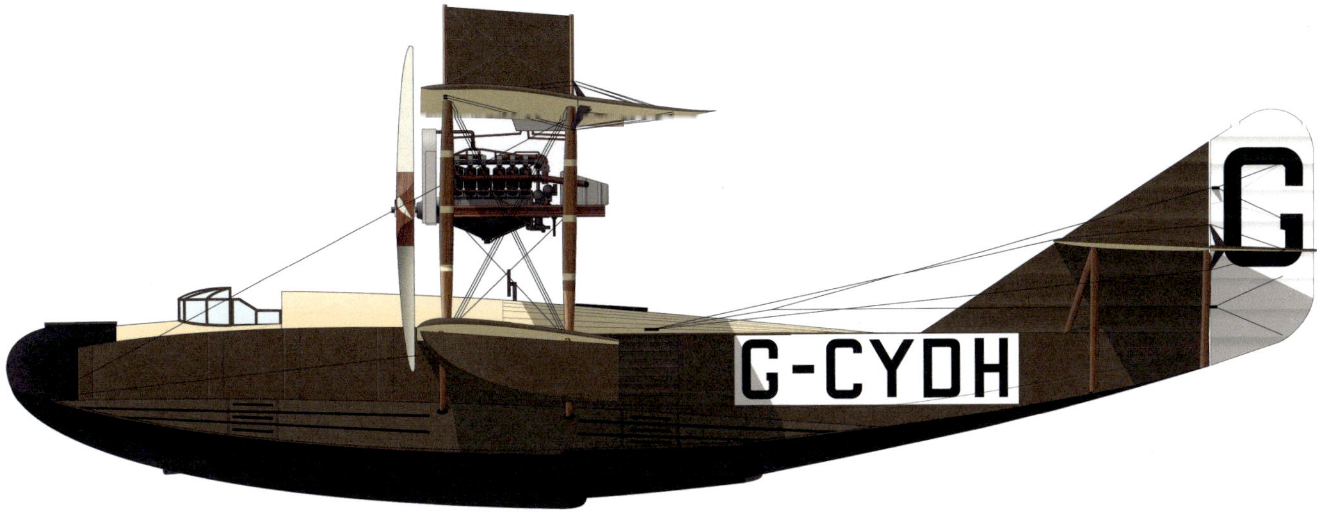

Felixstowe F.3 G-CYDH of the Canadian Air Board, 1921–1923. Canada received 11 F.3s in 1919 as a postwar gift. N4009 was registered G-CYDH on 11 July 1921 and was used to spot forest fires during the summer fire season. It was deleted 11 January 1923.

Appendix 2: The Development of the Large Flying Boat for Naval War Purposes, 1917–1919

By Captain A.E. Bolton
Copy in U.K. National Archives File AIR1/275/5
Transcribed and annotated by Colin A. Owers

Above: An F.3 beached. Note that the fins are a light colour on this boat.

INTRODUCTION

After the end of the First World War personnel were asked to write of their careers in the RAF. Some of these files are short and not very informative. Capt Bolton, on the other hand, presents his views on the work of the Felixstowe and Yarmouth flying boats of the RNAS/RAF over the North Sea in great detail and it is worthy of being presented in his own words.

The first large flying boats to be used in this country by the Royal Naval Air Service were of the Curtiss H.12 Large America Type, fitted with two Rolls-Royce engines, each of 250 H.P. They were assembled and fitted out at Felixstowe Air Station in 1916, under the supervision of Commander J. Porte and were first used for war purposes in the early months of 1917, when they met with immediate success as effective anti-submarine craft in the southern portion of the North Sea. They carried a crew of four, 1st Pilot, 2nd Pilot, Wireless Telegraph Operator and Engineer, and were armed with two Lewis Guns, one mounted in the forward cockpit, and the other in the aft cockpit.

These H.12 Flying boats carried in addition two 250 lb bombs, and were capable of a speed of about 80 knots.

The advent of the large flying boat marked the beginning of a new era in the naval side of aerial warfare, as they were a very distinct improvement over the large slow float type sea-planes at that time in both air speed, observational powers, bomb carrying, and radius of action.

Very splendid work was done against the German submarines in the region of the North Hinder Light Vessel, where it was the custom of the enemy to lie on the surface during the night, charging their batteries in preparation for the hazardous passage through the straights of Dover into the English Channel. Patrols were systematically carried out throughout the day over the whole area, and especially at dawn and sunset. Many enemy submarines were attacked and bombed and several sunk, and so harassing were the attacks of the flying boats that the Germans had practically to abandon their custom of running on the surface in the Southern part of the North Sea. In consequence, they were not able to negotiate with any degree of safety the passage of the Straights of Dover, and most of the German submarines based on the Ems and Wilhelmshaven had to travel to the English Channel and the south coast of Ireland via the north of Scotland.

This extra distance very considerably curtailed their destructive powers against the Allied Merchant Shipping, as owing to the extra distance – nearly 500 miles – the time they

Above: F.3 at Houton Bay on 25 May 1918. Probably a machine of No.306 Flight that remained independent after 1 April amalgamation of the RNAS into the RAF, The station name is prominently displayed on the hull that appears to be a light colour.

could cruise in the tracks of the shipping was considerably reduced.

This excellent work in the early days of flying boats certainly played a no mean part in the defeat of the submarine menace, and also resulted in the destruction of the German Zeppelin L.43 by Flt Lt B. Hobbs and Flt Sub Lt R.F.L. Dickey on the morning of 14 June 1917.

It was the many successes meet with by these early patrols that demonstrated to the Admiralty the possibilities of this new weapon at their disposal, and the production of flying boats, their development, and the training of pilots was expedited by every means in their power.

On the 13th April 1917 the first H.12 No.8666 was allotted to Yarmouth and piloted there by Fl Lt Leckie, who at that time had recently completed training as a Boat Pilot at Felixstowe. It was from that date that the Flying Boat as a reconnaissance machine can be said to have started. Every possible morning of favourable weather was seized upon by Fl Lt Leckie and Fl Commander C.J. Galpin, who acted as his second pilot, to cross to the North Dutch coast in search of Zeppelins. Many were attacked and escaped only by the superior climbing powers of the lighter-than-air to heavier-than-air craft, but on the morning of 14th May 1917 the German Zeppelin L22 was brought down and H.12 8666 had commenced her famous career of adventure, as, without a doubt, she was destined to become the most famous of war-flying boats, and earned for herself the honour, when her active service days were over, of a permanent place in the Imperial War Museum, London.[1]

The H.12 type flying boat, although a wonderful aerial weapon, had many disadvantages for war purposes which soon made themselves manifest. The first was her inability to rise from rough water without great risk of damage, as the design of the bottom was exceptionally weak, compared with the otherwise robust construction of the hull, especially at the point where the tail portion joined the step, and unless very great care was exercised in "taking off" and landing, there was a very big possibility of the hull breaking at this point. This was a very serious defect, and greatly depreciated the usefulness of the machine, more particularly at a station like Yarmouth, which did not enjoy the shelter of a harbour, and so the patrols possible were to a very great extent governed by the weather at the base – a non too reliable factor in the North Sea.

As a fighting machine the H.12 was under gunned for her size, and was at a distinct disadvantage when attacked from below, where her guns were both totally blind, and from most points behind the tail when the after gun could not be fired for fear of damage to the tail plane, rudder and elevator controls.

These difficulties were, however, almost totally overcome by the development under Commander Porte of the F.2.A type flying boat, a machine of all British design and construction. The hull of this boat was "V" shaped and curved from bow to

Above: A *Large America* low over the water displays the altitude adopted when in combat with German floatplanes.

stern, and the tail portion of the hull, therefore rose completely clear of the water when the boat was at rest. This served several useful purposes. Firstly, the boat could be landed with complete safety without fear of damage in much rougher seas than was possible with the H.12, as with this type of hull a "tail down" landing could be carried out at a much lower speed than the 'step' landing requiring a higher speed, and necessary with the H.12 type; reference to the following sketch will readily illustrate this detail.

*** Sketch of H.12 and F.2A in landing position *** – *Missing from file*. The 'V' shaped bottom also facilitated "taking off" in rougher waters, as it did not hammer the water whilst attaining flying speed to anything like the extent of the practically flat-bottomed H.12 hull.

In addition to the improvements in flying properties the "cocking up" of the tail in the "F" boat design made very practical the addition of two Lewis Guns normally housed inside the hull on the port and starboard sides respectively. Sliding doors on the sides of the boat just aft of the main planes uncovered ports, and these guns were so mounted on brackets that they could be swung outboard commanding a very large area of fire meeting at a point more than 20 feet astern. The possession of these guns made an attack from under the tail a very hazardous proposition for an enemy machine, whilst attack from above was also covered by a gun mounted on the top of the hull in the after cockpit. There was unfortunately still a blind spot immediately behind the tailplane but this was relatively small and was rarely found to give much trouble in attack from astern.[2]

The F.2A was very similar in overall dimensions to the H.12, the hull being 45 feet from the bow to the back of the rudder, and the span of the main planes 98 feet. It was fitted with two Rolls-Royce engines each of 350 H.P. and was capable of a maximum speed of eighty five to ninety knots. It carried 250 gallons of petrol sufficient for about eight hours duration at sixty five knots.

The armament consisted of six Lewis Guns, as in addition to the three after guns already mentioned, twin guns were mounted on a scarf ring in the forward cockpit, and one gun was superimposed on the cabin and manned by the 2nd pilot. The F.2A manned by a crew of five was undoubtedly the most formidable defensive fighting aircraft of the war as all guns were mounted on moveable mountings and on no other heavier-than-air machine was such a concentration of fire possible.[3]

At Yarmouth the fighting possibilities of the F.2A were most thoroughly investigated by the Boat Pilots and all the spare time possible was devoted to formation flying and the practice of tactical evolutions for the purpose of exploiting the gun power to the utmost advantage in all encounters with the enemy. Time after time the trouble taken in the perfecting of these manoeuvres was amply repaid in times of emergency and it reflects greatly to the credit of the Yarmouth crews that during the whole of the war no Yarmouth flying boat was ever shot down by the enemy on the Long Distance Reconnaissance

Patrols although in the encounters which occurred the Germans were nearly always in the heavy odds of at least two or three to one.

In accordance with the general practice of the British Admiralty throughout the war the Large Flying Boats and their activities were shrouded with as much secrecy as possible and even the general details of their construction rarely leaked out to any great extent. Even today not many people are conversant with the size and details of these flying boats so that a short description of the general arrangement of the interior of an F.2A may be of interest at this point.

In the bow was situated the forward cockpit in which was mounted on a Scarff ring the twin guns already mentioned. The height of the floor boards at this point was such that an average man was clear of the deck from about the waist upwards. On the starboard side in a convenient position just forward of the cockpit was the bomb sight, and the bomb release controls were fitted inside the hull where they could be conveniently reached by the observer whilst at using the sight.

About six feet further aft was the navigation cabin finding the 1st and 2nd pilot's seats, controls and instrument board. The 2nd pilot's seat was on the port side and so arranged on a hinge that his controls could be uncoupled and folded beneath his seat when not in use, and this permitted of access from the main portion of the hull to the forward cockpit.

The cabin consisted of an almost semi-circular framework of small glass windows with a roof of clear celluloid. The structure stopped just short of the seat and while affording complete protection to the pilot allowed the 2nd pilot when standing on the seat ample room to use his gun and also gave him an excellent position for observation in all directions.

In front of the 1st Pilot on the starboard side was the control wheel carried on a hooped shaped frame which moves in a fore and aft direction for the elevator controls whilst the wheel was rotated to port or starboard for the aileron or side balancing controls. The rudder was as usual in all aircraft worked by the pilot's feet. In front of the 1st Pilot also was the instrument board in the centre of which was a large vertical compass graduated every 2 degrees as on these long distance sea going machines accurate course steering was absolutely essential, even small deviations resulting in large errors at the end of a long course. Arranged around the compass in various convenient positions were the other instruments – the Altimeter showing the height in thousands and hundreds of feet above sea level – the Air Speed Indicator indicating the rate of the passage of the Boat through the air in knots – the Inclinometer giving the angle of rise and fall of the centre line of the Boat above and below the horizon – the Revolutions Indicators showing the number of revolutions per minute of the port and starboard engines respectively – the Oil Pressure Indicators showing the condition of the oil flow in the engines – and Thermometers recording the heat of the water in the engine radiators. The pilot at a glance was therefore able to gather all this necessary information concerning the performance of the Boat at any time. Reference to the Instrument board would not be complete without mention of that important little instrument commonly known to seaplane pilots as the Bubble: it was not much to look at being to all intents and purposes a horizontal spirit level but the glass tube instead of being straight was part of the circumference of a circle. As long as the Boat followed a straight path through the air or was banked correctly on a turn, that is so long as the Boat neither slipped inwards towards the earth or outwards away from it, the bubble remained in the centre of the glass. Any deviation from this position immediately indicated to the pilot that he was doing something wrong, either he was flying one wing down, or slipping sideways through the air. At all times the bubble was useful but when flying through fog or clouds when the sea was not visible, it was invaluable.

As regards the control of the engines the magneto switches were mounted on a small panel close to the right hand of the first pilot, and the throttle levers of the two engines were between the seats of the 1st and 2nd pilots so as to be of easy access to both.

Besides keeping a good lookout all round and relieving the 1st pilot of control when desired, it was the duty of the 2nd pilot to keep an accurate record of the movements of the Boat, navigational objects sighted and their bearing at that time, alterations to course, etc., and to calculate for the reference of the 1st pilot the necessary allowance to be made to the compass course to counteract the drift caused by the wind on the course to be followed. Special charts of convenient size were prepared by the Admiralty for use of the naval air pilot and to save calculation in the air as much as possible tables for wind allowance on various courses were made out and photographed down to small size.

Immediately aft of the pilots' seats the curve of the hull allowed the floorboards to be at (a) much lower level leaving about six feet of headroom inside the Boat. The Wireless Telegraph Operator sat at his instruments on the starboard side behind the 1st pilot and was therefore in an excellent position for giving signals to the pilots as they were received, and for taking signals for transmission to base.

Aft of the wireless operator came the petrol tanks, the bulkiest and heaviest equipment of the Boat. They consisted of five circular tanks each having a capacity of 50 gallons and spaced so as to allow of passage between them to the after cockpit. From these tanks the petrol was raised by a small wind driven pump to a 25 gallon tank in the top plane, from which the petrol flowed to the engines by gravity. It was the first duty of the Engineer to look after the petrol tanks and as each tank emptied, to switch over to a full one. Radiator thermometers were fitted conveniently for his information and a small semi-rotary pump and auxiliary water supply enabled him to see the temperature of the water in the engine radiators was not allowed to get dangerously high. In times of action the Engineer manned one of the after guns.

Behind the tanks was a short companion ladder to the aft cockpit hatch enabling easy access to the top after gun and in ordinary times the Gunlayer kept a good lookout aft

Above: N4534 lacks the "N" of the serial on the hull. This F.2A appears to have served at Calshot for its service life from May 1918 to at least January 1919.

from this position. In the sides of the cockpit were the sliding doors mentioned in the early part of this article concealing the remaining two after guns and which could be opened for both operational and fighting purposes.

From this aft cockpit to the tail plane the hull tapered rapidly away both in width and height and the space in consequence was useless for any practical purpose.

Some idea of the size of the F.2A Flying Boat will have been gathered from the foregoing description, and it may be of interest to note that the weight of an F.2A ready for patrol and complete with crew, petrol, guns, ammunition, batteries, and stores was just over five tons.

Precautions against forced landings at sea were taken as far as possible by a supply of compressed rations said to be sufficient to provide nourishment for five men for a week, and four gallons of fresh water were always carried. This practice followed the temporary loss of 8666 for three days from Yarmouth in September 1917, when after the landing it was unfortunately discovered that none of the crew had taken any food with them, and also they had only two gallons of fresh water on board.

Forced landings miles away from base on the mine strewn wastes of the North Sea where shipping was never seen during the war, were the worst enemy the naval air pilot had to fear, but when a seaplane was missing, the unfortunate crew had the consolation of knowing that no effort was being spared by their comrades to find them. Frail though the flying boats were when exposed to the anger of the North Sea, they were very much more seaworthy then the float type seaplanes, and throughout the war they were several more times subjected to gales whilst down on the water, there is no record of a flying boat actually foundering with her crew aboard. For these emergencies four sea trained carrier pigeons were carried in every boat and frequently found most useful. Other methods of getting into communication with base, such as by means of lifting the wireless aerial with a kite, were also successfully employed after forced landings. The greatest protection however against forced landings lay in the care and maintenance of the engines, and in her engineering staff Yarmouth was most fortunate; it speaks volumes of praise for their efficiency when from January 1918 to the Armistice sixty four thousand miles were patrolled by the Yarmouth Boat Squadron and only three machines during this time were forced to land with engine trouble.

Elsewhere will be found descriptions of the more interesting adventures of the Yarmouth Flying Boats such as the destruction of the Zeppelin L.22, the sighting and reporting of German Battle Cruisers, Light Cruisers, and Destroyers, the probable frustration of a raid on our shores thereby, and the fight off Terschelling Islands, but for the most part the work of the Flying Boat was chiefly more nerve wracking by its monotony than its excitement. Hour after hour roaring over the waters of the grey North Sea in fine weather and foul the crews of the Flying Boats had only the satisfaction of knowing their report was the better for being negative as the Admiralty

Above: N4402 a Phoenix F.3 at the factory's slipway at Bradford. This was an unlucky boat capsizing on its delivery flight to Dundee on 27 April 1918. Salvaged it was to be reconstructed. It was last listed in January 1919.

probably preferred to reports of enemy activities.

Commander Porte did not rest on his laurels of the F.2A, but pushed ahead with his construction and designs, and there is not the slightest doubt that had the war lasted the further developments of still larger, faster, and more powerful flying boats might easily have resulted in making it equally dangerous for Zeppelins to patrol in the northern part of the North Sea as it undoubtedly was for them in the southern portion, and our Fleet would have accordingly benefitted in relief from the enemy's vigilance in those waters.

The latest development of the F.2A at the end of the war was the Felixstowe Fury, a credit to British aerial supremacy. It was a large triplane of F.2A design weighing fifteen tons when ready for patrol, and driven by five Rolls-Royce engines each of 350 h.p. This enormous machine, in spite of its size was capable of attaining a speed of 120 knots and a height of ... feet, and carried sufficient petrol for a flight of ... hours, would have played a magnificent part in the conduct of naval warfare.[4]

Appendix 2 Endnotes

1. Whether 8666 was ever in the IWM's collection is unknown, but this great boat was not saved.
2. The side guns converged about 20 feet under the boat leaving only a small area where firing was not possible.
3. When on reconnaissance patrols and not carrying bombs the F Boats carried more guns for defence against the German seaplanes that waited for them. "Fighting tops" were proposed for the long range boats but were never used operationally.
4. The figures for altitude and duration are left blank on the original document.

Bibliography

Books

Abbatiello, J.J. *Anti-Submarine Warfare in World War I – British naval aviation and the defeat of the U-Boats,* Routledge, UK, 2006

Arthur, R.W. *Contact, Naval Aviators Assigned Numbers 1 to 2001,* Naval Aviation Register, DC, 1967.

Barnes, C.H. *Shorts Aircraft since 1900,* Putnam, UK, 1967.

Brew, A. *Sunbeam Aero Engines,* Airlife, UK, 1998.

Bruce, J.M. *British Aeroplanes 1914-1918,* Putnam, UK. 1957.

Bruce, J.M. *The Felixstowe F.2A,* Windsock datafile No.82. Albatros Productions, UK, 2000.

Carr, W.C. *Out of the Mists,* Hutchinson & Co Ltd, UK. (No date given but published during WWII).

Casey, L.S. *Curtiss- The Hammondsport Era 1907-1915,* Crown Publishers, NY, 1981.

Cole, C. & Chessman, F. *The Air Defence of Britain 1914–1918,* Putnam, UK, 1984.

Dodds, R. *The Brave Young Wings,* Canada's Wings Inc, Canada, 1980.

Goodall, M.H. *The Norman Thompson File,* Air-Britain, UK, 1995.

Grosz, P.M. *Brandenburg W.29,* Windsock Datafile No.55, Albatros Productions, UK, 1996.

Grosz, P.M. *Brandenburg W.12,* Windsock Datafile No.61, Albatros Productions, UK, 1997.

Gutman, J. *Naval Aces of World War 1,* Osprey, UK, 2012.

Pix (Hallam, T.D). *The Spider Web,* William Blackwood & Sons, 1919.

Henshaw, T. *The Sky Their Battlefield II,* Fetubi Books, UK, 2014

Herris, J. *German Seaplane Fighters of WWI,* Aeronaut Books, USA, 2012.

Hobson, C. *Airmen Died in the Great War 1914–1918 – The Roll of Honour of the British and Commonwealth Air Services of the First World War,* JB Hayward & Son, UK, 1995.

Imrie, A. *German Naval Air Service,* Arms & Armour Press, UK, 1989.

Jones, H.A. *The War in the Air,* Vol.2 to 6. The Claredon Press, Oxford.

Livock, G.E. *To the ends of the Air,* IWM/HMSO, UK. 1973.

Macmillan, N. *Tales of Two Air Wars,* G. Bell & Sons Ltd, UK, 1963.

Mikesh, R.C. & Abe, S. *Japanese Aircraft 1910–1941,* Putnam, UK, 1990.

Molson, K. "The Felixstowe F5L," *Cross & Cockade Great Britain,* Journal, Vol.9 No.2, 1978.

Moss, P. *The British Civil Aircraft Registers 1919–1928,* Air Britain, UK, 1969.

Nowarra, H.J. *Marine Aircraft of the 1914–1918 War,* Harleyford Publications, UK, 1966.

Owers, C.A. *The Curtiss H-12,* Albatros Productions, Windsock DataFile No. 125, Albatros Productions, UK, 2007.

Owers, C.A. *Hansa-Brandenburg Aircraft of WWI,* Vol.2 & 3. Aeronaut Press, USA, 2015.

Penrose, H. *British Aviation – The Great War & Armistice,* Putnam, UK, 1969.

Raleigh, Sir W. *The War in the Air,* Vol.1. Oxford Clarendon Press, UK. 1922.

Robertson, B. *Aircraft Camouflage and Markings 1907–1954, Harleyford* Publications Ltd, UK, 3rd revised impression, 1959

Robertson, B. *British Military Aircraft Serials 1912–1969,* Ian Allen, UK, 1969.

Robinson, A. "The Recognition of First War Flying Boats", *Cross & Cockade Great Britain,* Journal, Vol.11, No.1

Rossano, G.L. *Stalking the U-Boat, U.S. Naval Aviation in Europe During World War I,* University Press of Florida, USA, 2010.

Snowdon Gamble, C.F. *The Story of a North Sea Air Station,* Oxford University Press, 1928. Reprinted by Neville Spearman, UK, 1967.

Snowie. J.A. *Collishaw & Company – Canadians in the Royal Naval Air Service 1914–1918,* Nieuport Publishing Co, Bellingham, Canada, 2010.

Sturtivant, R. & Page, G. *Royal Navy Aircraft Serials and Units 1911-1919,* Air Britain Publication, UK, 1992

Taylor, H.A. *Fairey Aircraft since 1915,* Putnam, UK, 1974.

Wortman, M. *The Millionaire's Unit,* Public Affairs, NY. 2006.

Flying Officers of the US Navy 1917–1919, Reprint by Schiffer Publishing Ltd, USA, 1997.

Hansa-Brandenburg W.29, Wingnut Wings, NZ, downloadable from www.wingnutwings.com.

Hansa-Brandenburg W.12 (Early), Wingnut Wings, NZ, downloadable from www.wingnutwings.com.

Rigging Notes F.5 Boat Seaplane. Air Publication No.905, UK, 1922.

Technical Notes F.2A Flying Boat Large America, subtitled *Large America Class Flying Boat (Types F.2A and F.3) Rigging Notes.*

Articles

Bruce, J.M. "The Felixstowe Flying Boats," *Flight,* UK, from 2 to 23 Dec 1955.

Bruce, J.M. "The Felixstowe Flying Boats," *Aeroplane Monthly,* UK, October and November 1982. Reprinted in the *Cross & Cockade 25th Anniversary Seminar* papers, UK.

Hope, L.C. "Notes on Flying Boat Hulls," *Flight,* From 20 May 05 June 1920

Knapp, M.W. "Flying Boats over the North Sea," *Over the Front* Journal, USA, Vol.9 No.2, 1994.

Koch, D. "Aeromarine Airways – Its Aircraft and History," *Skyways,* USA, No.52, Oct 1999.

Moore, C.R. "Felixstowe F2A Flying Boats," *Scale Models,* UK, Dec 1977.

Nicolson, D. "Design and Construction of Flying Boats," *Flight.* From 3 to 10 July 1919.

Nicolson, D. "Flying Boat Construction," *Flight,* From 24 March to 01 April 1921.

Rennie, J.D. "Some Notes on the Design, Construction and Operation of Flying Boats," *Journal of the Royal Aeronautical Society,* Proceedings, 7th Meeting, 58th Session, 18.01.1923.

Shirley, N (Ed). "Wartime Memoirs of John Jay Schieffelin." *Over the Front* Journal, USA, Vol. 9 No.2, 1994.

Owers, C.A. "A Killingholme Diary," *Cross & Cockade, International,* UK, Vol.35, Nos.3 & 4, 2004.

Van Dorssen, H. & Gerdessen, F. "Interned! A list of aeroplanes of all nationalities that were interned by the Dutch authorities after landing in Holland during 1914–1918," *Cross & Cockade International,* UK, Vol.17, No.1.

Vicary, A. "High Visibility Colour Schemes on Felixstowe and Great Yarmouth Flying Boats," Cross & Cockade International, Vol.21, No.4.

Walpole, N. "Tresco's Sentinels," Flypast, UK, Dec 2008.

Documents from the RAF Museum, Hendon; the National Museum of Naval Aviation, Pensacola, Florida; the Australian, UK and US National Archives: the USN Heritage and History Command; Navy Yard, Washington, DC; the Museum of Flight, Seattle, Washington. The *Flight* online archive.

F-5-L 315 of the Brazilian Naval Air Service, mid-1920s. In 1923 Brazil acquired 14 F-5-Ls; these were given serials 1–5, 21–23, 311–316, and 321–326. They served until 1930. These colors are speculative.

F-5-L B-8 of the Argentinian Air Service, circa 1922. In 1922 Argentina purchased eight F-5-Ls and gave them serials B-1 through B-8. They served until 1931 as operational aircraft, then as trainers until 1935. These colors are speculative.

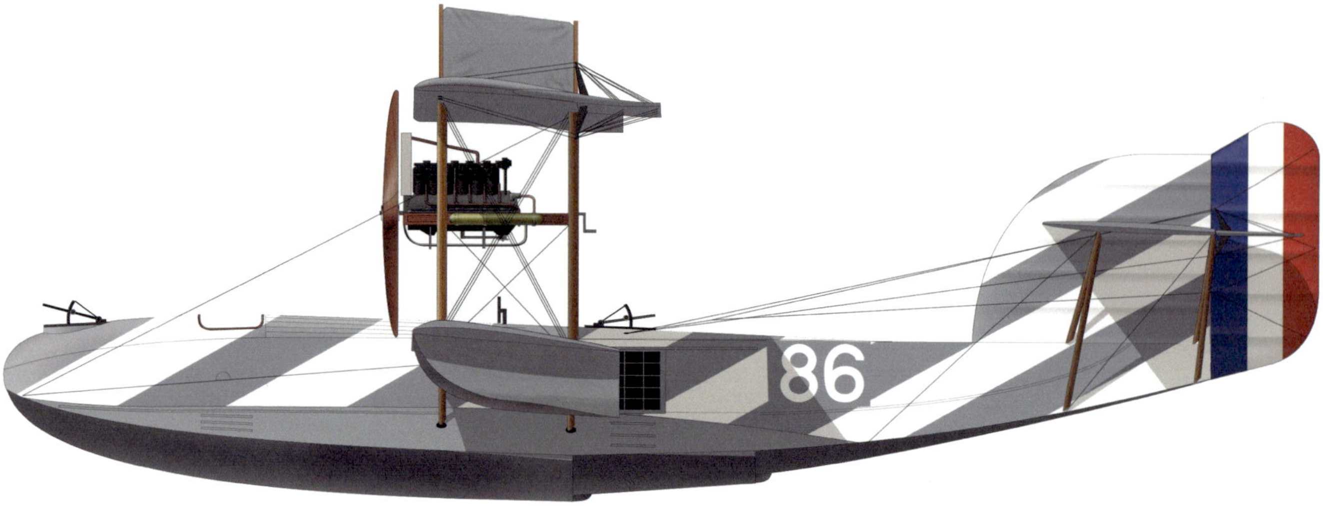

F-5-L of the U.S. Navy.

F-5-L A3565 of the U.S. Navy, circa 1922.

F-5-L of the U.S. Navy.

Aeromarine 75 *Santa Maria* of the Aeromarine West Indies Airways, circa 1921.

Aeromarine 75 *Buckeye* of Aeromarine Airways Inc., summer 1922.

Aeromarine 75 *Niña* of Aeromarine Airways Inc., 1922–1924.

Felixstowe F.5 N4490 assigned to Felixstowe in December 1918. N4490 was named *Aquila*.

Felixstowe F.5 N4490 operating from Malta in the 1920s; the wartime dazzle scheme has now been replaced. The photo of N4490 in flight appears to show it with a black and white hull like some other F.5s.

Felixstowe F.2A assigned to Great Yarmouth; pattern on lower hull and floats is speculative.

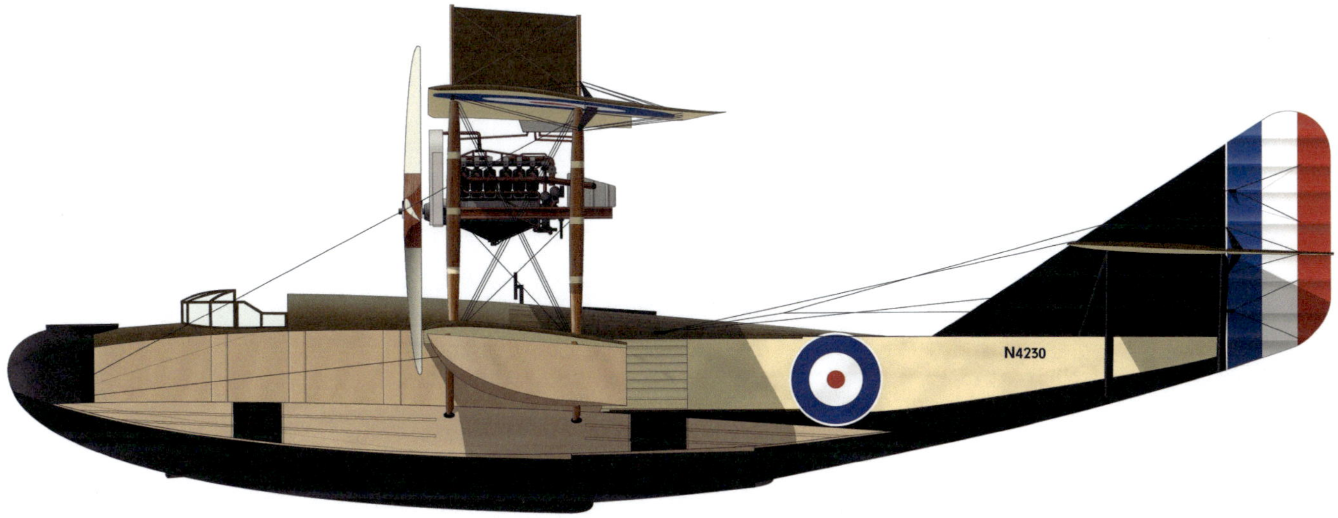

Felixstowe F.3 N4230 operating from Shoeburyness, May 1918. After undertaking hydrophone trails at Shoeburyness in May, the Dick, Kerr and Co. built N4230 dropped two 230 lb. bombs on a U-boat on 20 July. It was recorded at Houton Bay in January 1919.

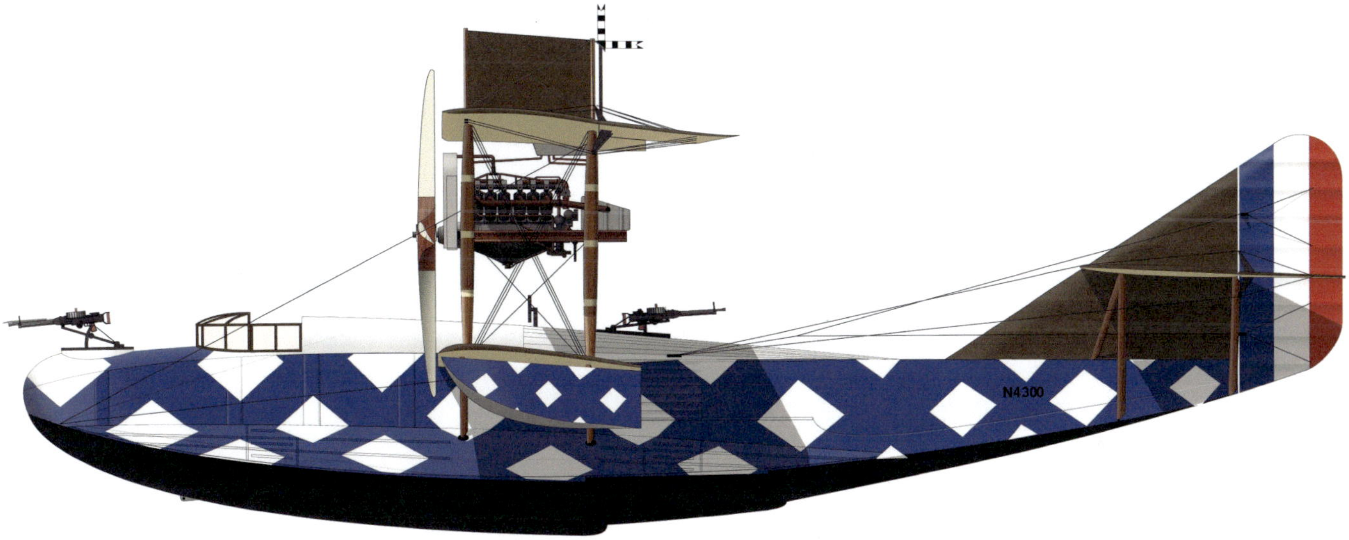

Felixstowe F.2A N4300 carried a semaphone device above the center section.

Felixstowe F.3 G-CYEN (ex N4015) of the Canadian Air Board, 1922. It crashed at Victoria Beach, Manitoba, on 8 September 1922 and was not repaired.

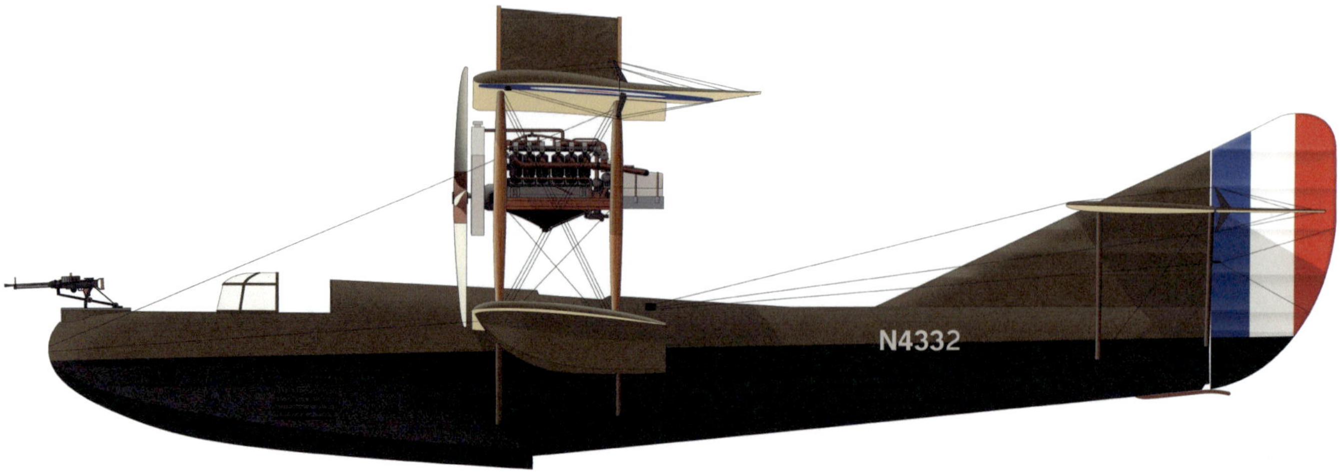

Curtiss H-12B N4332 was delivered in December 1917, assigned to Great Yarmouth in the spring of 1918, and deleted in May 1918. The dark lower hull is depicted as black but this is not confirmed.

F-5-L of the U.S. Navy based at Hampton Roads NAS postwar as indicated by the "HR-2" on the bow.

Curtiss H-16 *Sampaio Correia*, August 1922. In 1922 the New York World newspaper sponsored a flight from New York to Brazil by ex-US Navy pilot Walter Hinton. On 22 August 1922 Hinton misjudged a night landing in Guantanamo Bay, Cuba, wrecking *Sampaio Correia*. A replacement aircraft completed the trip.

Made in the USA
Las Vegas, NV
04 November 2026

57421426R10169